T0296871

The Applied Genomic Epidemiology Handbook

The Applied Genomic Epidemiology Handbook: A Practical Guide to Leveraging Pathogen Genomic Data in Public Health provides rationale, theory, and implementation guidance to help public health practitioners incorporate pathogen genomic data analysis into their investigations. During the SARS-CoV-2 pandemic, viral whole genome sequences were generated, analyzed, and shared at an unprecedented scale. This wealth of data posed both tremendous opportunities and challenges; the data could be used to support varied parts of the public health response but could be hard for much of the public health workforce to analyze and interpret, given a historical lack of experience working with pathogen genomic data.

This book addresses that gap. Structured into eight wide-ranging chapters, this book describes how the overlapping timescales of pathogen evolution and infection transmission enable exploration of epidemiologic dynamics from pathogen sequence data. Different approaches to sampling and genomic data inclusion are presented for different types of epidemiologic investigations. To support epidemiologists in diving into pathogen genomic data analysis, this book also introduces the analytic tools and approaches that are readily used in public health departments and presents case studies to show step-by-step how genomic data are used and evaluated in disease investigations.

Despite the breadth of scientific literature that uses pathogen genomic data to investigate disease dynamics, there remains little practical guidance to help applied epidemiologists build their ability to explore epidemiologic questions with pathogen genomic data. This handbook was written to serve as that guide. Including case studies, common methods, and software tools, this book will be of great interest to public health microbiologists or lab directors, bioinformaticians, epidemiologists, health officers, and academics, as well as students working in a public health context.

Chapman & Hall/CRC
Computational Biology Series

About the Series:

This series aims to capture new developments in computational biology, as well as high-quality work summarising or contributing to more established topics. Publishing a broad range of reference works, textbooks, and handbooks, the series is designed to appeal to students, researchers, and professionals in all areas of computational biology, including genomics, proteomics, and cancer computational biology, as well as interdisciplinary researchers involved in associated fields, such as bioinformatics and systems biology.

Stochastic Modelling for Systems Biology
Third Edition
Darren J. Wilkinson

Computational Genomics with R
Altuna Akalin, Bora Uyar, Vedran Franke, Jonathan Ronen

An Introduction to Computational Systems Biology: Systems-level Modelling of Cellular Networks
Karthik Raman

Virus Bioinformatics
Dmitrij Frishman, Manuela Marz

Multivariate Data Integration Using R: Methods and Applications with the mixOmics Package
Kim-Anh LeCao, Zoe Marie Welham

Bioinformatics
A Practical Guide to NCBI Databases and Sequence Alignments
Hamid D. Ismail

Data Integration, Manipulation and Visualization of Phylogenetic Trees
Guangchuang Yu

Bioinformatics Methods
From Omics to Next Generation Sequencing
Shili Lin, Denise Scholtens and Sujay Datta

Systems Medicine
Physiological Circuits and the Dynamics of Disease
Uri Alon

The Applied Genomic Epidemiology Handbook
A Practical Guide to Leveraging Pathogen Genomic Data in Public Health
Allison Black and Gytis Dudas

For more information about this series please visit: https://www.routledge.com/Chapman--HallCRC-Computational-Biology-Series/book-series/CRCCBS

The Applied Genomic Epidemiology Handbook

A Practical Guide to Leveraging Pathogen Genomic Data in Public Health

Allison Black and Gytis Dudas

CRC Press
Taylor & Francis Group
Boca Raton London New York

CRC Press is an imprint of the
Taylor & Francis Group, an **informa** business
A CHAPMAN & HALL BOOK

Designed cover image: Gytis Dudas

First edition published 2024
by CRC Press
2385 NW Executive Center Drive, Suite 320, Boca Raton FL 33431

and by CRC Press
4 Park Square, Milton Park, Abingdon, Oxon, OX14 4RN

CRC Press is an imprint of Taylor & Francis Group, LLC

© 2024 Allison Black and Gytis Dudas

ISBN: 978-1-032-53029-1 (hbk)
ISBN: 978-1-032-53026-0 (pbk)
ISBN: 978-1-003-40980-9 (ebk)

DOI: 10.1201/ 9781003409809

Typeset in Latin Modern font
by KnowledgeWorks Global Ltd.

For the mentors we've had,
And the mentors we hope to be.

Contents

Preface

THIS BOOK WAS WRITTEN LARGELY during 2021–2022, a pandemic period where SARS-CoV-2 brought massive changes to our lives and our work, and the desire to share and use pathogen genomic data for public health action greatly accelerated. As we worked – not just using sequence data to understand SARS-CoV-2 epidemiology but also teaching public health officials how to interpret epidemiological dynamics from those data – we identified a significant gap in the genomic epidemiology literature. There are excellent review papers providing short, accessible summaries of how pathogen genomics can support aims in public health (e.g., see Armstrong and colleagues [1]). On the other end of the spectrum are scientific papers describing the methods and findings of genomic epidemiological studies in technical detail. We felt that something was missing from the middle; a resource that introduced the basic theory and utility of genomic epidemiology, with a focus on how to use those data in public health practice, and provided step-by-step guidance for using pathogen genomic data in epidemiologic investigations. This book is our attempt to fill that gap.

It is our intention that after reading this handbook, you should be able to:

- Understand how genomic epidemiology can support certain investigations;

- Design genomic surveillance data collection to meet your needs; and

- Apply genomic epidemiology to routine investigations in public health practice;

Given this book's narrow focus on using pathogen genomic data in applied epidemiology, there are many things that this book will not be.

It is not a review of the primary literature in genomic epidemiology, nor do we aim to provide an exhaustive description of all the questions that scientists can investigate with genomic epidemiology. We will not present every method for genomic epidemiological analysis, nor provide information on the entire suite of available analytic tools. Rather, this handbook is meant as a practical guide to applied genomic epidemiology. As such, we focus on the questions that we see public health practitioners encounter most frequently, and present analytical methods and tools that are easily used within public health departments and other applied epidemiology settings.

For whom is this handbook written?

This book aims to help you understand how to use pathogen genomic data for public health surveillance, outbreak response, and public health decision-making. This book is for you if you are already involved in, or want to develop a program for, pathogen genomic data collection, pathogen genomic data analysis, pathogen genomic data interpretation, and/or policy evaluation in public health. For example:

- Public health microbiologists or lab directors developing a genomic surveillance program.

- Bioinformaticians working in public health, and wanting to increase their familiarity with the goals, theory, and approaches specific to genomic epidemiology.

- Epidemiologists who typically work with surveillance data, but who want to integrate molecular information into the investigations they conduct.

- Health officers or other policy makers who want to understand more about pathogen genomic data as a source of epidemiological information.

- Academics collaborating with public health institutions who want to learn more about genomic epidemiology and how these methods support the standard questions we ask in public health.

How should you read this handbook?

You can think of this resource as funnel-shaped, moving from imperative concepts for all readers towards more specific implementation information that is most pertinent for those readers specifically involved in data collection and analysis. We recommend that all readers read the first two chapters introducing the utility of pathogen genomics in public health and the fundamental theory underlying genomic epidemiology. These two sections will help you understand how genomic data enrich public health investigations and the basic mechanics behind genomic epidemiology. These sections will also introduce the common language you'll encounter when discussing genomic epidemiology studies.

Readers involved in designing or implementing genomic surveillance and epidemiology programs within their agencies should also read the chapters introducing sampling strategies and outlining the broad use cases for genomic epidemiology. These sections will help describe when you can use pathogen genomic data and how you should approach collecting it.

From there, readers who wish to see investigations falling under these different use cases in action should peruse the provided case studies. Our intent with the case studies is to show step-by-step why we initiated an investigation, how we framed our question of interest, and how we investigated it, including quality control, evaluating competing hypotheses, and weighing uncertainty. While narratives presented in the published literature are by design cohesive and smooth, with these case studies we aim to show exactly how an investigation occurred, including bumps and questions along the way.

To help guide readers into hands-on analysis of pathogen genomic data, Chapter 6 introduces "Tools and Methods for Applied Genomic Epidemiological Analysis", which lives towards the end of this handbook since it is primarily pertinent to those readers actively involved in genomic epidemiological analysis. We are well aware that this chapter will go out of date the quickest; however, we feel that the tangible support that this chapter provides for getting started warrants its inclusion.

Finally, this book concludes with a deeper dive into theory and analysis, touching on the greater genomic complexity that exists for both viruses and bacteria beyond what we introduced in the first theory chapter. Some concepts in these chapters may veer slightly towards the academic. However, we've encountered enough situations where this additional knowledge was useful that we wanted it to be available to readers.

While you might skip these chapters initially, you will likely come back to them if you dig more deeply into genomic epidemiology.

We've read enough textbooks to know that reading textbooks generally isn't "fun". But we do sincerely hope that this resource helps you begin to integrate pathogen genomic data into your public health work, wherever you're starting from. And we hope that the fun will come from getting to explore this new territory.

About the Authors

Allison Black is a genomic epidemiologist passionate about strengthening public health's ability to use pathogen genomic data for outbreak response, disease surveillance, and decision-making. While she trained as an academic genomic epidemiologist, she now leads molecular epidemiology activities at the Washington State Department of Health.

Gytis Dudas is an evolutionary biologist at Vilnius University's Life Sciences Center working on RNA virus evolution at both long- and short-term scales. Most of Gytis' PhD and postdoctoral work focused on the genomic epidemiology of Ebola virus in West Africa and MERS coronavirus in the Arabian Peninsula.

Contributors

Allison Black
Washington Department of Health
Seattle, Washington

Gytis Dudas
Vilnius University
Vilnius, Lithuania

Taj Azarian
University of Central Florida
Orlando, Florida

Janeece Pourroy
Chan Zuckerberg Initiative
San Francisco, California

CHAPTER 1

The Value of Pathogen Genomics in Applied Epidemiology

Allison Black

Washington State Department of Health, Seattle, Washington

Gytis Dudas

Vilnius University, Vilnius, Lithuania

H AVE YOU EVER WONDERED "What would be the benefit of using pathogen genomic data in this investigation?" This chapter attempts to answer that question. We'll describe the utility of pathogen genomic data for public health surveillance and for outbreak response. We distinguish between surveillance and outbreak response use cases for two reasons. Firstly, the turnaround times for which genomic data retain utility are different for surveillance versus outbreak response. Typically, outbreak response requires sequencing and genomic analysis to occur more rapidly. Secondly, the sampling frames that you use for surveillance are different from what you would use for outbreak response. We will elaborate on sampling frames in Chapter 3. We recognise that getting a genomic epidemiology program up and running can be slow and challenging. We therefore also discuss how genomic data can remain useful even when they are not available in real time. This chapter should be pertinent to most readers, since it provides a broad rationale for why genomic surveillance and genomic epidemiology help support applied epidemiology activities.

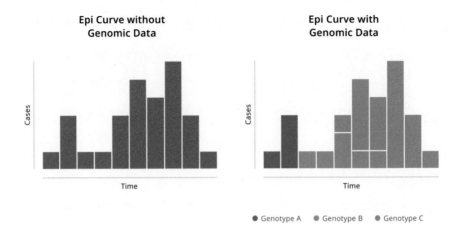

Figure 1.1 Epidemiologic curves without and with the addition of genomic data. On the left, we see the epidemiologic curve without genomic information. Given just the shape of the curve, we might infer that this outbreak started with a single introduction event with some degree of sustained transmission. On the right, we see the same epi curve, but with cases coloured according to their genotype. The addition of genomic data suggests that, in fact, this outbreak is attributable to three distinct introduction events of divergent genotypes and that these different genotypes contributed in different degrees to the overall outbreak.

1.1 THE VALUE OF GENOMIC EPIDEMIOLOGY FOR SURVEILLANCE

Genomic data provide additional resolution to determine relationships between cases. From a surveillance standpoint, this ability to delineate clusters of related cases more sensitively allows public health practitioners to see separate pathogen lineages circulating within a particular area of interest. This capacity allows the epidemiologist to see distinct chains of transmission even when they circulate in a population concurrently, a finding that can be challenging to see in case surveillance data alone (Figure 1.1).

What is the benefit of detecting these distinct transmission chains? Firstly, this additional layer of resolution enables the public health practitioner to detect the emergence or introduction of new lineages into their community or surveillance area, and distinguish introduction events from endemic transmission within the surveillance area. Developing a more precise understanding of how the processes of introduction and endemic transmission drive incidence enables public health authorities to intervene in more precise and tailored ways. For example, if most cases within the surveillance area result from multiple introductions of distinct lineages that transmit only minimally after introduction, then policies that seek to reduce the frequency with which introductions occur are likely to be more effective in reducing case counts. In contrast, if most cases are attributable to a single circulating transmission chain, then public health policies focusing on reducing travel-acquired cases are likely to be less effective, and greater focus should be oriented towards interventions designed to interrupt transmission within the community.

Taking our ability to distinguish between distinct transmission chains a bit further, when we have genomic surveillance data collected over time, these data can help the epidemiologist to see how different clusters contribute to the overall disease burden (Figure 1.2). For example, perhaps you detect four circulating genotypes within your community. Despite the existence of these four transmission chains, the majority of the cases in your community can be attributed to the circulation of just one of those transmission chains. In this case, the public health practitioner can focus attention on understanding what risk factors or demographic profiles appear to be most associated with the primary circulating transmission chain, with the hope that a more detailed understanding of that particular transmission chain will underscore more tailored and effective interventions.

Furthermore, evaluating the contribution of different pathogen genotypes to case counts over time can provide more precise situational awareness about how different variables shape the epidemiology of a pathogen. For example, we may be concerned about how travel and holiday behaviour contribute to cross-border pathogen spread. Genomic surveillance data can enable us to investigate that question, monitoring the prevalence of particular genotypes in different countries and observing changes in genotype-specific incidence rates following major travel periods. Such analyses enabled scientists to monitor the rise of the B.1.177 lineage of SARS-CoV-2 viruses in Europe during the summer of 2020, attributing the rising dominance of the variant to holiday travel, rather than to increased viral fitness or transmissibility [2].

While most mutations are simply typos in the genome and have little or no effect on the pathogen, some mutations can change the dynamics of transmission. Such lineages, which we may refer to as being "more

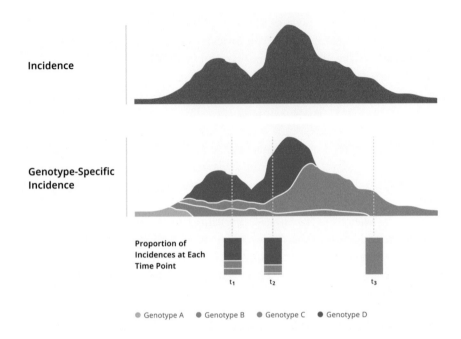

Figure 1.2 Toy example of monitoring genotype frequencies. In the top panel of this figure, we show a theoretical incidence plot. Below, the same incidence plot is shown given the addition of genomic data, which now allows us to measure genotype-specific incidence. Having a genotype-specific measure of incidence allows us to see how the frequency of different genotypes changes over time. Importantly, that information can provide critical situational awareness about an outbreak or provide warnings of genotypes that we may need to monitor more closely.

fit", do emerge. By monitoring genotype-specific incidence rates, we are well poised to detect an acceleration in the transmission of a particular lineage. Enabled with high-resolution descriptive epidemiological information, we can follow up with analytical epidemiological studies to test hypotheses about why a particular strain is rising in frequency. This process was critical in detecting the emergence of SARS-CoV-2 lineage B.1.1.7 (Alpha in the World Health Organization nomenclature) in the United Kingdom, and attributing the rise in frequency to the actual increased transmissibility of B.1.1.7 lineage viruses [3, 4].

Finally, a change in commonly reported symptoms or illness severity associated with an increase in a particular genotype could indicate a change in the clinical presentation of illness, possibly associated with a genotypic change in the pathogen. Like standard surveillance data, genomic data are observational data, and therefore to infer causality between genotype and presentation you would need to draw upon other experimental approaches.

1.2 THE VALUE OF GENOMIC EPIDEMIOLOGY FOR OUTBREAK RESPONSE

Pathogen genome sequences can help us detect or rule out linkages between cases. Similarly to how specific case definitions improve the sensitivity of traditional epidemiologic study designs, genomic data can support outbreak response by improving one's ability to accurately classify which cases form an outbreak cluster. Below are some toy examples to clarify how this works, and to demonstrate how this information is useful in guiding public health practice.

In the first toy example, imagine that three skilled nursing facilities (SNFs) in your community are all experiencing outbreaks of a particular illness. As spaces of congregate living that typically house more vulnerable people, such facilities may be more prone to a higher incidence of disease. Thus, you may ask yourself, are the outbreaks I see in these three facilities independent outbreaks, attributable simply to the higher risk setting? Or might these outbreaks be linked? Furthermore, are cases detected within each of the facilities linked? Or do some cases form a cluster, while others have simply been detected due to enhanced screening?

Figure 1.3 shows how genomic data can help elucidate some of these possibilities. From the surveillance data, we can see cases detected across all three SNFs. When we sequence the viruses from those infections, an interesting picture unfolds. Each SNF has a subset of cases with closely related genome sequences (indicated here by colour). Each SNF also has cases whose genome sequences are distinct from each other and from the related infections found across all of the SNFs. This theoretical picture helps us understand a few things. Firstly, these are not three independent outbreaks, but rather one outbreak across three facilities. This may cause us to ask whether residents are transferred between facilities regularly, or if staff members work at multiple facilities. Secondly, not all of the cases that were detected across the SNFs appear to form part of the outbreak. Rather, we have one primary cluster, along with detections of

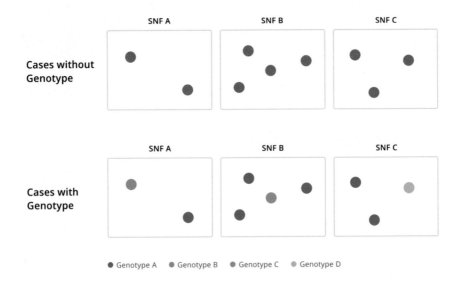

Figure 1.3 Cases of a disease across three skilled nursing facilities (SNFs), without and with genomic sequencing data. In the first row, we see that we have multiple cases of disease across three separate institutions. Without any additional knowledge about the cases, we might conclude that each SNF is experiencing an independent outbreak. The addition of genomic sequence data suggests that these SNFs are in fact all part of the same outbreak, given that they each have cases that are infected with the same genotype of the disease. Furthermore, the sequence data allows us to see which cases are really part of this multi-facility outbreak, and which infections are in fact unrelated prevalent cases detected through outbreak response efforts.

infections that were likely acquired in the community, separately from the outbreak.

Similarly, genomic data can distinguish cases that are not linked, even if they are detected within the same setting or facility, or infect individuals with similar demographic profiles. Distinguishing between unrelated and related cases within a single setting can enable epidemiologists to more clearly see what precise factor(s) is driving transmission.

Let us take another toy example, in this case a factory where you detected an elevated incidence of disease above baseline. The cases are distributed across the factory, among both managerial staff who

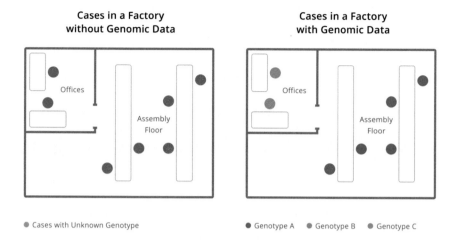

Figure 1.4 Here, it is shown that how the addition of genomic data changes our understanding of transmission within a factory. On the left, we see seven cases of infection, all within the same factory. This may lead us to believe that all of these cases are related, and perhaps became infected at work. The addition of genomic data helps us resolve this picture more. Once we have genotype information, we see that cases that work on the assembly floor all appear related, while cases among individuals who work in the offices appear unrelated, and are likely prevalent cases detected through enhanced surveillance efforts. Determining that workers in the office are not part of the factory outbreak helps us to see that the occupational transmission risk appears related to work on the assembly line. This would allow us to target our intervention efforts towards this group of workers.

primarily work in offices and individuals working on assembling products. All the cases have been detected around the same time period, meaning that the temporal pattern supports a single outbreak across the entire facility. In that case, what factor is facilitating transmission? How will you act in order to interrupt transmission?

Figure 1.4 shows how the addition of genomic data enables you to see that the cases among office-based managerial staff are unrelated to the infections occurring amongst workers on the assembly line. While all cases occurred among employees of the factory, infections among managerial staff are genetically distinct from cases that occurred among assembly staff. Therefore, managerial infections appear to have been acquired externally from the workplace. In contrast, all cases among assembly staff are genetically related. Distinguishing the office staff cases from the assembly staff cases allows you to see that any transmission occurring within the factory appears to be specific to the assembly area of the factory. This improves the sensitivity with which you can detect the factors or behaviours that are facilitating transmission, and hopefully making it easier to intervene.

1.3   THE VALUE OF RETROSPECTIVE DATA

In public health, we frequently want short turnaround times that allow us to learn and act as quickly as possible. However, initially you might have delays in generating and using pathogen genomic sequence data. This is completely normal and goes hand-in-hand with the challenges of building out new microbiological capacity within the public health lab, new data infrastructure for performing bioinformatic analysis and data linkage, and new inferential tools for analysing the data. While decreasing turn-around-times will help you to utilise the data in real-time to inform outbreak response, genomic data that are retrospective in nature are still useful!

For example, sequencing retrospective samples can increase or change our understanding of the dynamics and timing of outbreaks. You may find that sequencing older samples retrospectively identifies earlier detections of particular pathogen lineages, therefore suggesting that a lineage might be more prevalent in the present than assumed. Such updated information may improve models used for epidemiological forecasting, or can guide policies surrounding therapeutics administration, if it is known that different lineages show different susceptibilities to those therapeutics. Similarly, hypotheses about the role of particular events in initiating transmission or disseminating an outbreak might be altered or dismissed based on retrospective data. An early sequence might also become a "missing link" that adds weight towards a hypothesis that two distinguishable local transmission chains arose locally and were not the result of two separate introduction events. While it is rare for retrospective samples to be paradigm-shifting, we will look at two examples where retrospective sequencing and phylogenetic analysis changed our fundamental understanding of a disease's epidemiology.

The human immunodeficiency virus type 1 (HIV-1) group M pandemic resulted from a spillover event from Central African chimpanzees,

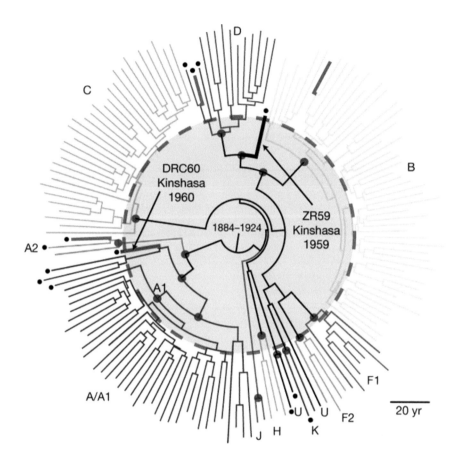

Figure 1.5 Phylogenetic tree of HIV-1 sequences from Worobey and colleagues [5]. Copyright © 2008, Macmillan Publishers Limited. All rights reserved.

and the Democratic Republic of Congo (DRC) was likely a starting point. However, owing to the late identification of human immunodeficiency virus (HIV) as the causative agent, the earliest genetic sequences of HIV-1 came from samples collected in the 1980s, likely multiple decades after the initial spillover event from chimpanzees to humans. With many years of transmission and evolution occurring before viruses were ever sequenced, the genome sequences from the 1980s are already genetically diverse. Retrospective sequencing of HIV genomes from two blood samples collected in Kinshasa, the capital of the DRC, in 1959 and 1960 were illuminating. Both samples belonged to different subtypes of HIV-1 group M and thus indicated that substantial genetic diversity of HIV-1 was already present in the DRC by 1960 (Figure 1.5). This finding helped

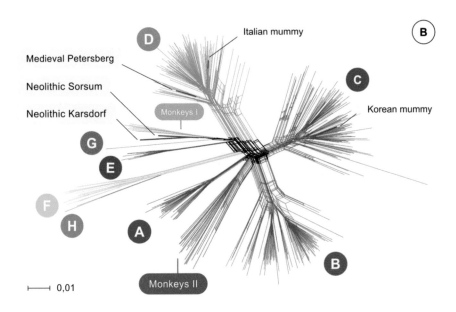

Figure 1.6 Phylogenetic network of hepatitis B virus sequences from Krause–Kyora and colleagues [6]. Copyright © 2018, Krause-Kyora et al.

to revise our understanding of the origins of the HIV-1 pandemic, pushing back our estimates of when HIV-1 group M viruses spilled over into the human population towards the turn of the century, a period of time involving rapid urbanisation in Central Africa.

A similar and even more exaggerated example comes from ancient hepatitis B virus samples. Hepatitis B virus (HBV) is a circular double-stranded DNA virus that infects hepatocytes and causes hepatitis. A number of genotypes of hepatitis B have been described, and genetic diversity tends to correspond to broad geographic regions. Since hepatitis B virus has a DNA genome and double-stranded DNA is relatively stable, it has been possible to extract DNA from well-preserved remains that are hundreds to thousands of years old, from all around the world. This work has allowed us to sequence ancient HBV viruses and compare them to contemporary sequences. One of the most surprising findings of these studies has been that all historical hepatitis B virus sequences to date clearly fall within known contemporaneous genotypic hepatitis B diversity (Figure 1.6), indicating that the diversification of hepatitis B virus predates any of the ancient infections.

In both the HIV-1 group M and hepatitis B virus studies, retrospective sequencing provided a lot of new information because no information existed prior to contemporary sequences. Though this information isn't actionable for any response happening now, it does nonetheless paint a much clearer picture of when these viruses emerged, and puts constraints on hypotheses about past events. Even though we have explored examples with timescales on the order of decades to hundreds of years, the broader lessons would hold in a similar contemporary situation.

Even when data from an epidemic are plentiful, we may still have local blindspots. One example of this has been the Guinean portion of the 2014 West African Ebola virus epidemic, where a seemingly rare viral lineage (marked GN-1 in Figure 1.7) was detected by three different groups who were sequencing sporadically during the epidemic. Each time the lineage was detected, it seemed to circulate at such low prevalence that each set of investigators thought that the lineage would likely not persist beyond the study period. The persistence of this lineage is remarkable because it was only ever detected at low frequencies and primarily in Guinea, despite the presence of a fairly robust genomic surveillance system across the region. A potential hypothesis arising from these data is that this lineage actually circulated at higher frequencies, allowing the lineage to be sustained over long periods, but that it circulated in an area within Guinea where case detection or sequencing coverage was not as dense. When the lineage reached areas with good sequencing coverage, it was observed, but only rarely. If we assume that PCR testing was done at the source location and that those samples were somehow preserved, retrospective sequencing in regions of Guinea where sequencing was less intense could illuminate the circumstances that allowed this lineage to circulate for so long while only rarely being detected.

At the end of the day, situations where retrospective sequencing can significantly alter our perspective will be more common in resource limited settings, and we expect that retrospective sequencing generally will confirm transmission that we hypothesised was occurring.

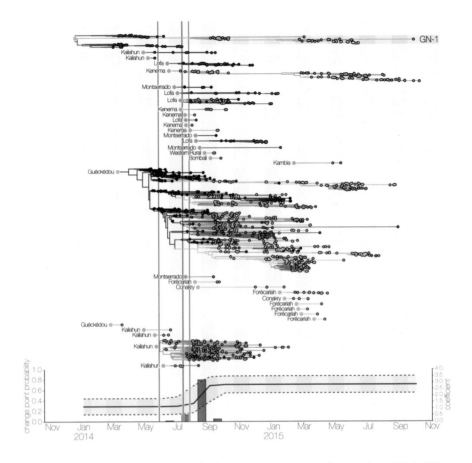

Figure 1.7 Phylogeny of all Ebola virus genomes from the 2014 West African epidemic split into continuous periods of evolution at country level (green is Guinea, blue is Sierra Leone, red is Liberia) from [7]. The lineage marked GN-1 (at the top) had been detected by three separate groups sequencing in Guinea at different times and appeared rare or on the verge of extinction each time. The long persistence of an apparently rare and geographically confined lineage is unusual, so retrospective sequence data could be very illuminating here. Copyright © 2017, Macmillan Publishers Limited, part of Springer Nature. All rights reserved.

CHAPTER 2

Fundamental Theory in Genomic Epidemiology

Allison Black

Washington State Department of Health, Seattle, Washington

Gytis Dudas

Vilnius University, Vilnius, Lithuania

ONE OF THE STRENGTHS of incorporating genomic data into epidemiological investigations is that it provides an additional, independent data stream by which to assess infectious disease dynamics. One of the challenges that comes hand-in-hand with that strength is that genomic epidemiology uses theory, analytical approaches, and jargon that surveillance epidemiologists may not be familiar with. In this chapter, we introduce the fundamental theory that underlies genomic epidemiological analysis and describe the terminology that genomic epidemiologists frequently use in discussing and interpreting our analyses. This chapter, which summarises the principles and mechanics of genomic epidemiology, should be pertinent to most readers of this handbook.

2.1 THE OVERLAPPING TIMESCALES OF PATHOGEN EVOLUTION AND PATHOGEN TRANSMISSION

Our ability to explore infectious disease dynamics using evolutionary analysis of pathogen genome sequences depends on a fundamental principle; pathogens evolve on roughly the same timescales as they circulate through a population of hosts. This principle means that the evolutionary trajectories of pathogens are shaped by the kinds of epidemiological and immunological forces that we, as public health practitioners, want to

DOI: 10.1201/9781003409809-2

learn about. Pathogen genetic diversity becomes distributed in different ways depending on varying host movements, transmission dynamics, environments, and selective pressures, among other forces. Exploring those patterns can help us to understand to what extent these different factors shape epidemics.

In the following sections, we will describe how mutations occur within a single infected individual and discuss how this leads to viral diversity observed at the population level, across multiple infected individuals in an outbreak.

2.1.1 Viral Diversity Accumulates Over the Course of a Single Individual's Infection

In thinking about how pathogen diversity accumulates over an epidemic, we begin with the processes occurring during a single individual's infection. In the example that follows, we will consider an RNA virus. These pathogens are highly amenable to genomic epidemiology because they evolve rapidly and replicate to large population sizes. This enables us to visualise and discuss the dynamics of the evolutionary processes more readily.

To begin, imagine the index case of our theoretical viral outbreak. Upon infection, the virus enters into that person's cells, hijacking some of the host cell machinery to make the proteins needed to generate progeny virions. The virus must also copy it's RNA genome such that these genome copies can be packaged into the progeny virions. To do this, RNA viruses rely on a protein called the RNA-dependent RNA polymerase (RdRp). While our cells have polymerases that transcribe DNA to RNA, these won't work for making an RNA copy of an RNA genome. Thus, almost all RNA viruses bring along their own RdRp for performing the task of replicating their genome.

Unlike our polymerases, the majority of RdRps lack proofreading capabilities. That means that during the process of genome replication, when the polymerase makes a mistake and incorporates the wrong nucleotide into the new genome it is transcribing, that base will stay there uncorrected, representing a change in the genome sequence of the "child" (the newly copied genome) compared to that of the "parent" (the template genome). These replication "typos" happen frequently; broadly speaking, one or two mutations like this occur *every single replication cycle* [8,9]. With a random change occurring at one site in the genome pretty much every replication cycle, a genome length on the order of tens of thousands of sites, and millions or even billions of progeny virions being generated over the course of a single person's infection, pretty much every single change to a genome sequence that could occur will occur during a single person's infection. Thus, within a single infection, there is a large amount of pathogen genetic diversity, which we typically refer to as **within-host diversity**.

2.1.2 Stochasticity and Selection Influence Variant Frequency Within an Infection

It can be easy to think that mutations that yield all this within-host viral diversity represent changes the virus is making *towards* some trait, but this is not so. These changes in the nucleotide sequence are simply transcription errors, like typos that you might make while sending off a rapid email, and they will be distributed across all of the sites of the genome. Some of these mutations will have detrimental impacts, even lethal ones, that make a progeny virion less fit or even unviable. We refer to such mutations as **deleterious mutations**. Some mutations will have absolutely no effect on the fitness of the virion at all; we refer to these as **neutral mutations**. Some mutations could confer a fitness benefit to the progeny virion; these are **beneficial mutations**.

While the occurrence of mutations themselves is a random process governed primarily by the error rate of the RdRp, the impact that different mutations have on the ability of the progeny virion to infect cells and replicate will influence the frequency of those mutations within the diversity of an individual's infection [10]. Mutations that are lethal or highly deleterious will be purged from the viral population quite quickly, as the progeny virions that carry those mutations fail to complete their replication cycle. Conversely, if a mutation is beneficial, perhaps it allows the virion to replicate more quickly, then the virion carrying that mutation will generate greater numbers of progeny, and the frequency or that mutation in the viral population will rise. Neutral mutations, which do not result in any functional changes to the virus, will rise or fall in frequency stochastically. Notably, deleterious to neutral to beneficial is a spectrum, and how significantly the mutation will change in frequency depends on how impactful the mutation is, as well as chance.

2.1.3 When a Transmission Event Occurs, the Within-Host Viral Diversity of the Infector is Sampled and Transmitted to the Recipient

Over the course of a single individual's infection, mutations occur, generating slightly genetically different viral populations within the infected person. The frequency of these different variant populations can rise, fall, or stay the same, as governed by chance and selective forces [10]. But in genomic epidemiology, our interest is typically in the process of transmission *between* individuals. So now we must discuss what happens with this within-host diversity at the time of transmission.

At the time of transmission, the within-host viral diversity of the infector is sampled and transmitted to the infectee. How many virions are "sampled" from the infector, and how many virions it actually takes to cause an infection in the recipient, varies from pathogen to pathogen. This concept is termed the **transmission bottleneck**. When transmission bottlenecks are narrow, just a small fraction of the viral diversity present within the infector is passed along to the recipient to initiate their infection. When transmission bottlenecks are wide, many virions are sampled from the infector and transmitted to the recipient.

While you do not need to know the width of the transmission bottleneck to conduct genomic epidemiological analysis, we bring up this concept because the width of transmission bottlenecks influence how genetic diversity gets passed along between individuals over the course of an outbreak. When the bottleneck is narrow, and just a small fraction of within-host diversity is passed along, random chance will play a significant role in which virions are transmitted and initiate the recipient individual's infection [11]. In contrast, when the transmission bottleneck is wide, then the sample of the infector's within-host diversity is more likely to mirror the extent and frequency of their within-host diversity, and the sample of viruses that founds the recipient's infection is likely to be similar to the infector's viral diversity at the time of transmission. Thus, the width of the transmission bottleneck influences how the viral diversity accumulating within a single infection gets passed along.

Because the within-host diversity of an infection is changing over the whole duration of that person's infection, transmission events occurring at different time points in the infector's infection will result in different samples of the viral diversity being transmitted. Similarly, because only a sample of the infector's viral diversity is transmitted on to a recipient, if a single infector were to infect many recipients at the same time, those recipient infections would be founded with slightly different samples of the infector's within-host diversity as well.

2.1.4 Consensus Genomes Provide a Summary of the Within-Host Diversity

Despite the importance of within-host diversity to the overlapping timescales of pathogen evolution and transmission, in genomic epidemiology we most often look at a summary of within-host diversity, not the entirety of the diversity. This summary is the **consensus genome**. The consensus genome represents the most frequently observed nucleotide at each site in the genome at the time of sample collection. At some sites in the genome there may be very little within-host diversity, and the vast majority of sequencing reads will support the same nucleotide. At other sites, there might be higher levels of nucleotide diversity. In such cases, this diversity can be summarised in the consensus genome with a **nucleotide ambiguity code** (https://droog.gs.washington.edu/parc/ images/iupac.html) (Figure 2.1). These codes are letters that are not A, C, T, or G, but denote what nucleotide mixture was observed. For example, if at a site in the genome you have 40% of sequencing reads supporting an A and 60% of sequencing reads supporting a C, you might choose to summarise this diversity in your consensus genome by using the ambiguous site M, which means A *or* C. There are various decisions surrounding what threshold you use for making an unambiguous or ambiguous base call. However, a deep discussion of these trade-offs, and the parameterisation of bioinformatic pipelines for calling consensus genomes, is beyond the scope of this introduction.

While the consensus genome sequence provides a summary of the within-host viral diversity at a cross-section in time, it does not capture the full course of the within-host diversity. Moreover, the consensus genome does not capture all of the diversity present at the time of sampling. Many of the mutations that arise during the process of viral replication will remain at such low frequencies that a "real" mutation is not discernible from a mutation arising from PCR amplification or sequencing errors. As such, the consensus genome will not capture many of the mutations that occur over the course of an infection. This is one of the reasons why consensus genome sequences can be identical between closely linked infections, even though viruses are mutating with every replication cycle. This dynamic also means that the rate at which we

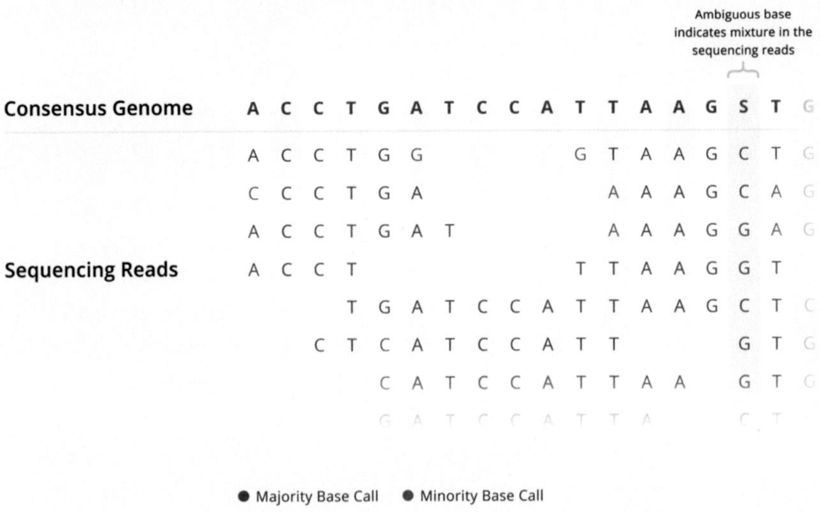

Figure 2.1 This plot provides an example of how a consensus genome is called given sequencing read data. Multiple short sequencing reads are stacked together according to which portion of the genome they cover. The consensus genome nucleotide call is typically the most common base observed at a site across all of the sequencing reads. When different sequencing reads support different nucleotide calls at a specific site, we can assign an ambiguity code within the consensus genome to capture that mixture.

observe changes in the consensus genome sequence of different infections is slower than the biologically governed mutational rate of the pathogen. Because the rate at which we observe these changes is different from the actual fundamental mutation rate of a pathogen, we have varied terminology for describing these processes, which we describe next.

2.2 TERMINOLOGY FOR DESCRIBING CHANGES IN GENETIC SEQUENCES

Vocabulary can be tricky, and the meanings of different terms may vary between academic domains. Furthermore, you may hear multiple terms for describing the same phenomenon. Sometimes these terms are synonyms, and other times they may have distinct meanings. Here, we aim to provide some clarity surrounding terms used in genomic epidemiology to discuss observed changes in microbial genetic sequences.

When dealing with microbial populations, the terms **mutation** and Single Nucleotide Polymorphism (**SNP**) are often used interchangeably. They refer to changes in the genetic sequence of the organism, at a single site in that organism's genome. You can observe a mutation or a SNP by comparing multiple aligned genetic sequences to each other. You may also hear mutations referred to as **alleles**, although this term can be confusing since it has a slightly different meaning when discussing the genetics of organisms that only carry one copy of their genetic material (such as viruses and bacteria) versus the genetics of organisms that carry two or more copies of their genetic material (such as humans).

The pattern of mutations that you observe across a sequence, summarised by the consensus genome sequence, is typically described as the **genotype**. Because viruses and bacteria are haploid, meaning they only carry a single copy of their genetic material, this single sequence defines their genotype. When you observe multiple samples with identical consensus genome sequences, we often describe these as multiple detections of the same genotype.

We use the term **substitution** to denote when a mutation has become completely dominant, and that all sequences in a particular population now carry that mutation. When this population-wide replacement happens, that mutation is said to be **fixed** in the population. Knowing when a mutation has fixed, and therefore become a substitution, is challenging since it requires knowledge about the genetic diversity of the entire population. Therefore, while the term is used quite frequently, it may not always be used entirely correctly. You will likely encounter the term "substitution" most frequently when discussing the rate at which we expect to observe nucleotide changes accumulating; these rates are frequently referred to as **substitution rates**, although given the challenge of truly knowing when a mutation has become a substitution, they are most appropriately called **evolutionary rates**. We discuss these various rates, and how to measure them, in the next section of this chapter.

2.3 MUTATION RATES, EVOLUTIONARY RATES, AND THE MOLECULAR CLOCK

The **mutation rate** denotes the actual rate at which a microbe's DNA or RNA polymerase makes errors while replicating the genome. This quantity is challenging to measure, since it requires specialised experimental designs. Thus, the actual intrinsic mutation rates of many organisms are not known.

Most mutations that the polymerase makes while replicating the genome are so detrimental that the organisms that carry them die out quickly, and those mutations are never observed. Thus, much of the genetic variation we actually observe is neutral or nearly neutral. These neutral and nearly neutral mutations accumulate in a population in a way that depends on the population's size (more replicating individuals means more chances for new mutations to arise) and the intrinsic mutation rate of the organism.

This rate at which mutations accumulate after selection has filtered out deleterious variation is called the **evolutionary rate**. Unlike the mutation rate, which depends most greatly on the functional characteristics of the polymerase, the evolutionary rate is the product of multiple intersecting features of the organism. These include the mutation rate, the amount of time it takes for the organism to replicate its genome and produce progeny, the size of the population of replicating individuals, the ability of the organism's genome to tolerate mutations, and the degree and type of selection acting upon the organism [10]. Beyond these organism-specific factors, the evolutionary rate will also be influenced by your sampling scheme [12]. For example, a dataset containing sequences collected infrequently over longer time periods may have a slower evolutionary rate, since there has been more time for purifying selection to remove deleterious variation from the population. Conversely, densely sequenced outbreaks may show slightly higher rates of evolution, since the intensive sampling of infections over short time frames may capture more of the deleterious or mildly-deleterious variation that would otherwise be purged from the population.

While the probability of a mutation occurring during viral replication is random, when we look through time at large populations of organisms, we see that mutations accrue across the genome according to the evolutionary rate. This signal of evolution through time is often referred to as the **molecular clock**, because the number of mutations that have accrued tells us something about the amount of time that has passed. While the terms "molecular clock" and "evolutionary rate" may be used interchangeably, the molecular clock is the principle that allows us to translate between nucleotide changes and calendar time, and the evolutionary rate represents the speed at which the clock ticks. Molecular clocks enable genomic epidemiologists to translate an observed number of mutations into an estimate of how much calendar time was necessary for that variation to accrue, allowing us to explore disease dynamics along a familiar timescale.

The simplest way to estimate the evolutionary rate is to look at the correlation between when a sample was collected and the genetic divergence of the sequence compared to the root of the tree. The root of the tree represents the ancestor of all sequences in our dataset. In most genomic epidemiological studies, the root is not a sample that we have sequenced, but rather a sequence that we infer during the process of making a phylogenetic tree. It represents an ancestor that would have existed and circulated some time ago given the pattern of sequence diversity that we have sampled and observed.

If we take the inferred genetic sequence of the root, and compare it to the observed sequence of a sample in our tree, we can count up how many mutations differ between the two sequences. We term this measurement of genetic divergence the **root-to-tip distance**. The root-to-tip distance of a sample acts as the y-coordinate in the plot that we are building up. The x-coordinates are given by the sample collection date (Figure 2.2).

Using all of the sequenced samples in our tree, we build up a scatter plot, which in this context is often referred to as a **clock plot** or a **root-to-tip plot** (Figure 2.3).

To improve accuracy and precision, we estimate the evolutionary rate using many sequences sampled serially, that is, over time. Then, to estimate the evolutionary rate, we draw a regression line through all the points. The slope of this line provides an expectation of the evolutionary rate (Figure 2.4). This estimate, which is drawn from the subset of samples that we include in our dataset, is specific to the sequences we are considering and how they were sampled. As such, you may see evolutionary rates shift slightly when analysing different datasets of the same pathogen, but usually this variation is minimal.

Evolutionary rates are frequently given in one of two forms; normalised to the genome length, or non-normalised. Normalised evolutionary rates represent the evolutionary rate of *a single* site in the genome. These normalised rates are useful because they can be compared across different organisms. In contrast, non-normalised evolutionary rates represent the total number of nucleotide changes you would likely observe across the entire genome, over a period of time (typically one year). They can be more interpretable, but they are specific to the pathogen and its genome size. When using non-normalised rates, it is important to remember that two different pathogens with identical *per site*

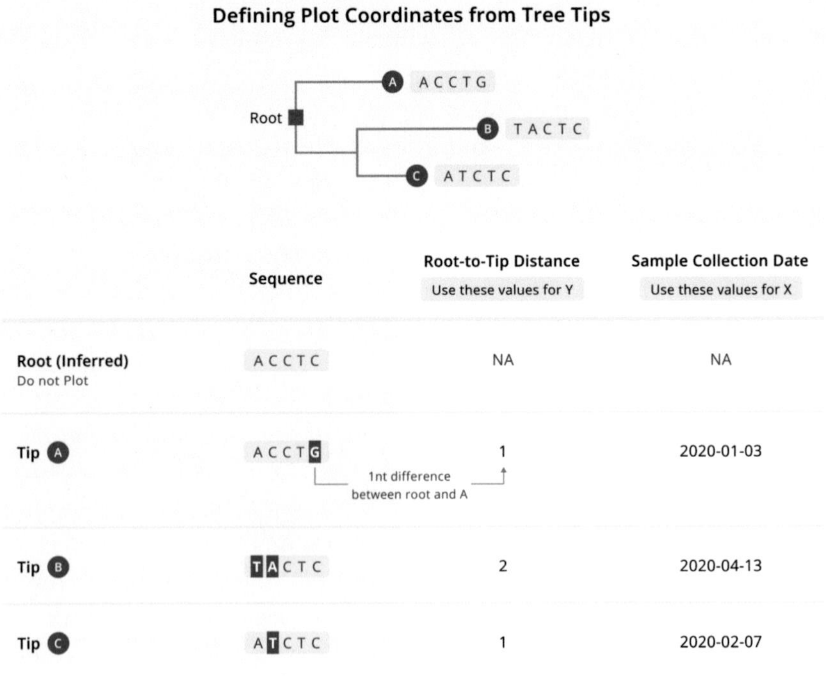

Defining Plot Coordinates from Tree Tips

	Sequence	Root-to-Tip Distance Use these values for Y	Sample Collection Date Use these values for X
Root (Inferred) Do not Plot	A C C T C	NA	NA
Tip A	A C C T G 1nt difference between root and A	1	2020-01-03
Tip B	T A C T C	2	2020-04-13
Tip C	A T C T C	1	2020-02-07

Figure 2.2 Here is a theoretical phylogenetic tree, for which we will cal-
culate the root-to-tip distances for the three sequenced samples: A, B,
and C. The sequences of A, B, and C are indicated, as is the inferred
sequence of the root of the tree. The number of mutations differing between the root sequence and each of the sampled sequences gives the
root-to-tip distance, which we use as the y-value for the sequenced sample. The x-value comes from the date when that particular sequenced
sample was collected.

evolutionary rates, but different genome lengths, will accumulate different total numbers of mutations over time; the pathogen with the
longer genome will accumulate more mutations. We emphasise this point
since we have invariably seen alarming headlines about rates of evolution that compare normalised to non-normalised rates, or that compare
non-normalised rates between different pathogens with different genome
sizes.

Luckily, translating between normalised and non-normalised evolutionary rates is simple. If you start with the normalised, per-site

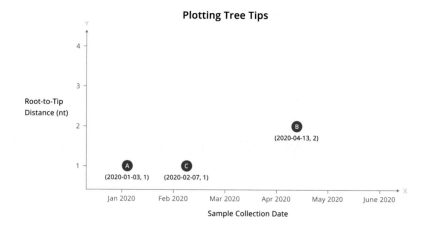

Figure 2.3 Using the coordinates that we inferred from the tree in the previous figure, we can create our root-to-tip plot. For the purposes of root-to-tip plots, we only perform this process for the sampled tips of the tree. We do not plot points for the root or the other internal nodes in the tree.

evolutionary rate, and multiply that quantity by the genome length, you will arrive at the non-normalised rate. Similarly, dividing a non-normalised genome rate by the genome length will return the normalised, per-site evolutionary rate.

How should you interpret the evolutionary rate? Generally, we describe this quantity as an expectation of the amount of nucleotide divergence we would observe between two randomly selected sequences sampled some amount of time apart from each other. For example, imagine you estimate that your pathogen of interest has a non-normalised evolutionary rate of 20.5 substitutions per year. This rate means that, on average, if you randomly select two sequences from your population that were collected exactly one year apart, they would be separated by 20.5 mutations. While in reality if you looked at the number of nucleotide mutations separating sequences sampled a year apart you would observe various whole numbers of mutations, if you did this procedure many times over, you should arrive at the expectation given by the slope of the regression line through your root-to-tip plot.

This ability to translate observed genetic divergence into an estimate of how much calendar time has passed opens up a number of tools for us.

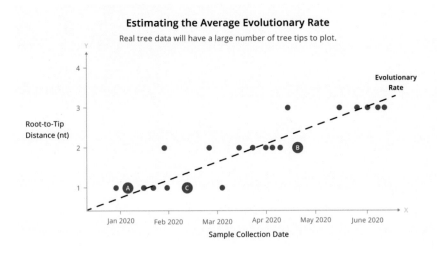

Figure 2.4 Here, it is shown what a true root-to-tip plot would look like for a tree with many sequences sampled longitudinally. Note that the coordinates that we inferred above are still plotted as above. We are using the same procedure for getting and plotting coordinates, we've simply added more data, which comes from a larger tree with more sequenced tips. Our estimate of the evolutionary rate of this dataset comes from the slope of the regression line that we fit through these points. Thus having more sequenced samples, and having sequenced samples collected over longer time frames, helps improve the precision and accuracy of our evolutionary rate estimates.

Firstly, it enables "back-of-the-envelope" calculations of how much time likely separates two sequences. This can be particularly handy when investigating potential epidemiological linkage between two sequenced cases. Secondly, molecular clocks enable us to translate genetic divergence trees into temporally resolved phylogenetic trees, or time trees. Finally, root-to-tip plots are also excellent (and quick!) ways to perform data quality control and look for interesting biology and epidemiological patterns. For example, if sequencing or bioinformatic assembly has gone awry, your sequences will often show greater nucleotide divergence from the root than you would expect given when they were sampled. They would thus appear as outliers that deviate from your molecular clock. Similarly, if the date of collection is recorded incorrectly, then a sequence might look over- or under-diverged given the sampling date.

This might be hard to identify in a spreadsheet, but will often jump out at you in a root-to-tip plot. Finally, deviation from the molecular clock may also alert you to more rare infectious disease dynamics, such as relapses or reactivations of latent disease. Indeed, observations of deviation from the molecular clock provided some of the first evidence that Ebola virus could be sexually transmitted from recovered individuals after many months [13, 14].

2.4 MULTIPLE SEQUENCE ALIGNMENT

Multiple sequence alignments (MSAs) are the input data type for any phylogenetic analysis in genomic epidemiology. Before widespread sequence data were available, phylogenies were inferred from character matrices that collated different measurements or trait presence/absence for various organisms. Each row in this matrix corresponds to an organism/species/taxon, and each column is a trait that has been measured or observed, for example, "Does this organism fly? What is its average weight? What is its average lifespan?" Based on the sharing of these character states, we can hypothesise common ancestry for groups of organisms, especially if multiple traits appear to support it (e.g., lactation, fur, dentition, etc. for mammals).

Multiple sequence alignments function exactly the same as these character matrices – each row is still an organism in question but now instead of tens or hundreds of trait measurements we have thousands or tens of thousands of nucleotides comprising their genomes as the columns. This comes at a cost, however, since we need to determine which columns (nucleotides at each site) of one organism go with which columns of the others. The job of multiple sequence alignment algorithms is therefore to provide a hypothesis of common descent for each position between any number of sequences. A misalignment of two molecular sequences is equivalent to shifting rows in our morphological trait matrix described above, where after the shift we might accidentally instruct the phylogenetic inference algorithm to place a fish that now errantly lactates and flies onto a phylogeny amongst vertebrates that perform these functions as well. The worst thing is that the phylogenetic algorithm will do this if thus instructed.

Alignment algorithms will attempt to align any sequences you give it, even if they are entirely random and unrelated. Similarly, phylogenetic inference algorithms will infer a phylogenetic tree no matter how high or poor the quality of the underlying alignment. Of course the vast majority of the time you will be dealing with outbreaks of limited genetic diversity, or pathogens with close relatives where alignments will be unambiguous, but it is highly recommended to inspect alignments by eye, with highlighting of sites that are different from a reference or consensus of the entire alignment. Sections of sequences that are misaligned, or contain sequencing or assembly errors, will be very easy to spot this way. When performing genomic epidemiological analysis, we recommend adopting a habit of inspecting your multiple sequence alignments every time you see particularly long branches or sequences that seem out of place in the phylogenetic tree.

2.5 PHYLOGENETIC TREES

2.5.1 What is a Phylogenetic Tree?

Phylogenetic trees are hierarchical diagrams that describe relationships between organisms. Specifically, phylogenetic trees, or phylogenies, are hypotheses of common descent, whereby sequences that have seemingly inherited the same mutations will cluster together.

Phylogenetic trees are composed of **tips** (also called **leaves**), **internal nodes**, and **branches**. Tips represent directly observed samples; you know the genetic sequence of the tips of a tree because you actually sequenced the samples. These are the samples that you use to infer the phylogenetic tree. In a tree, tips can be presented as a branch that simply ends, or more commonly, the name of the sample or a shape (often a circle) will be placed at the end of the branch to indicate the tip. The x-axis position of a tip is given by the observed genetic divergence between that sample and the root sequence of the tree.

Internal nodes represent hypothetical common ancestors that were not directly observed, but that we infer existed given the genetic patterns we see among the tips. These two types of objects are connected by branches. In genetic divergence phylogenies, when the pathogen population is sampled very densely it is possible to find samples (tips) with zero branch length. That is, the tips are at the same divergence level as their inferred common ancestor. This indicates that the genotype of the inferred common ancestor is the same as an actual sequence in the dataset.

Branches form the connections between nodes and tips in the tree, and they represent direct ancestor-descendant relationships. Branches can be **external** if a branch connects an internal node to its descendent tip or **internal** if a branch connects two internal nodes. Since phylogenetic trees model a hypothesis of genetic descent, internal nodes and tips have a sequence (either inferred or known, respectively). Mutations occur along branches, such that when the ancestor (or parent) node does not have a particular mutation, but the descendant (or child) does, then the mutation was gained along the branch that connects parent to the child.

When evolution proceeds via this process of descent, mutations that happen early on, and are inherited as part of the genetic backdrop that new mutations continually arise upon, will be present across many of the sampled and inferred sequences. It is this pattern of shared mutations and unique mutations that enable the hierarchical clustering that occurs in a phylogenetic tree (Figure 2.5). Groups of sequences that share a particular mutation are inferred to have inherited that mutation and cluster together, descending from the branch where that mutation occurred. In contrast, sequences that do not share that same mutation are likely not part of the same pattern of descent, and will cluster in a different part of the tree. Taking all of these patterns together allows us to build trees in which smaller and smaller groups of sequences cluster together with a shared pattern of inherited mutations. While most of the structure of the tree is formed by looking at patterns of shared genetic variation, some mutations will be unique to a sampled sequence. In this case, these mutations occur on the external branches that lead to the tips. The external branch length will be a function of the number of unique mutations that tip has.

Within the phylogenetic tree, subgroups encompassing a common ancestor and all of its descendants are typically referred to as **clades** (or sometimes **lineages**) (Figure 2.6). We define clades according to the mutations that the samples grouping together within the clade have inherited and all share. Clades are useful to consider when thinking about potentially altered biological properties. If a particular lineage has acquired novel mutations, and those mutations are passed on to its descendants, the entire clade would be expected to show the particular functional characteristics associated with the mutation(s) they all share.

Notably, because the phylogenetic clustering pattern is hierarchical, so are clades. Two sequences might cluster together into a small clade defined by a mutation that only they share. But those sequences can also be grouped within a larger clade, with additional samples, defined by a different set of mutations that occurred further back in the tree and were inherited by additional samples in the tree.

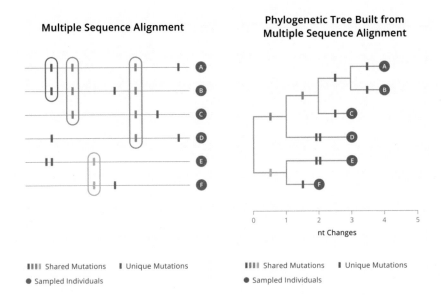

Figure 2.5 On the left is a theoretical multiple sequence alignment of genomes A through F. Shared and unique mutations are found in multiple samples. We use this pattern of shared and unique mutations to build the phylogenetic tree, which hierarchically clusters tips according to which mutations they share. Mutations occur along branches, such that tips that descend from a branch will share that mutation. When mutations are shared by more samples, then those mutations would have occurred more deeply in the tree. Mutations that are unique to samples occur on external branches, whose only descendent is the sampled tip.

It is common to encounter other embellishments on phylogenetic trees. For example, branches can be coloured to indicate inferred traits of ancestors in the tree. Common traits reconstructed on trees include geography or host. This annotation helps genomic epidemiologists understand the history and dynamics driving an epidemic. Another common addition, particularly to phylogenetic tree figures in scientific papers, are numbers or markers placed near or on nodes. These markers are a common way to indicate the statistical support for a given node (remember that each node is only a hypothesis of common descent). It has also become popular to mark mutations, particularly those changing amino acids in proteins, above the branch along which the mutations likely occurred.

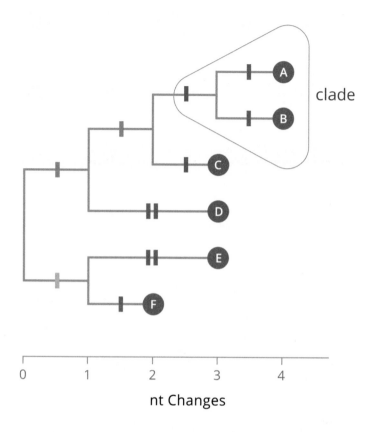

Figure 2.6 A hypothetical phylogenetic tree with a clade shown. The clade containing samples A and B is defined by the blue mutation that they both share, and that differentiates them from other samples in the tree.

Phylogenetic trees come in two categories: rooted and unrooted. An unrooted tree simply displays the inferred relationships between the samples without making any assumptions about where the tree begins. In contrast, rooted phylogenies explicitly state our hypothesis regarding the directionality of evolution. Within genomic epidemiology, you will primarily encounter rooted phylogenies, which is why we will concentrate on them in this section.

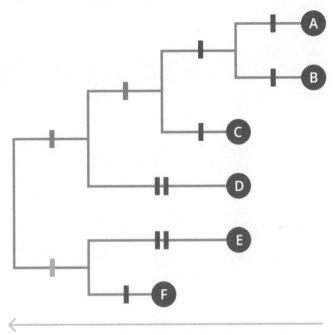

Basal - the direction of going deeper into the tree away from the tips.

Figure 2.7 A hypothetical phylogenetic tree showing directionality. When we refer to a node being more basal within the tree, we mean deeper in the tree, closer to the root. The root is the most basal internal node in the tree. This phylogenetic tree is rectangular and oriented left-to-right. This means that the left-most internal node is the most basal, and the root, and that evolution proceeds forward from that node, moving from left-to-right.

   Within a rooted phylogeny, the direction of evolution proceeds from the root out towards the tips. Nodes that are **basal** in the tree, or deeper in the tree, are ancestral to nodes that are closer towards the tips (Figure 2.7). Rooted phylogenies are typically displayed in rectangular form, with the direction of evolution proceeding from the left to the right. In this format, the most basal node, the root, is the node that is furthest to the left, and descendants of that node will appear to the right of that node. That said, rooted phylogenies can be shown in other conformations as well, such as circular or top-to-bottom. In these cases, just remember that the direction of evolution proceeds from the root toward the tips. The branches are scaled in terms of genetic divergence, or the number of expected changes per nucleotide site.

2.5.2   Assessing and Reading a Phylogenetic Tree

Phylogenetic trees might seem simple, but correctly interpreting them is not always straightforward. For beginners and advanced users alike, interpreting a phylogenetic tree can present a challenge because the y-axis in a phylogenetic tree is meaningless. It is used to lay out all of the tips so they do not overlap – the proximity of any two tips along the y-axis does not indicate anything about their relatedness. This also means that the branching in a phylogeny can be rotated without altering the meaning of the diagram in any way. Because phylogenies sacrifice an entire dimension for clearly laying out tips, there is also a limit on how much information can be packed into simple phylogenetic tree figures, and even moderately sized phylogenetic trees presented under ideal conditions can vary enormously in their ability to communicate key messages.

In learning how to assess a phylogenetic tree, we will consider only rooted phylogenetic trees, since you probably will not encounter scenarios where unrooted trees are necessary. One of the first things to check in a rooted tree is whether the root of the tree seems appropriate. The root of the tree represents the common ancestor of all your samples. Rooting determines the order of branching in the tree, and thus whether the root is correct or not will make all the difference when interpreting the tree. Ask yourself: "What lies on either side of the very first split in the tree?" You should be able to find a reference sequence representing the earliest known genome of a pathogen, or a fairly distinct but related organism (often termed an **outgroup**), or some known historical split in the pathogen's population on one side of the root. Of course there are always caveats to this rule, but generally the rooting and directionality you observe in the tree should match your understanding of the pathogen's descriptive epidemiology. Figure 2.8 shows an example of how incorrectly rooting a phylogenetic tree can fundamentally, and incorrectly, alter your understanding of an evolutionary trajectory.

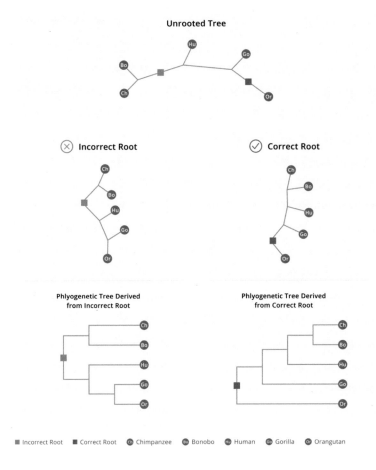

Figure 2.8 Rooting the phylogenetic tree of great apes. At the top, we see the unrooted phylogeny, as well as two possible places we might place the root, one of which is correct and the other is incorrect. In selecting the root, we describe a pattern of evolutionary descent out from the root. When the incorrect root is selected, on one side of the split we have chimpanzees and bonobos. On the other side of the split we have humans, gorillas, and orangutans. This does not match our understanding of evolutionary relationships between humans and great apes, and thus this particular split should alert us that the tree is incorrectly rooted. However, when the correct root is selected, as shown on the right hand side, we see a pattern of descent in which orangutans are on one side of the split, and chimpanzees, bonobos, humans, and gorillas are on the other side of the split. Chimpanzees and bonobos are most closely related to each other, followed by humans, then by gorillas.

Emanating from the root are branches representing direct lines of descent from the common ancestor. When looking at a genetic divergence phylogenetic tree, you will typically see that branch lengths are scaled in terms of a normalised number of substitutions *per site in the genome*. For increased interpretability, branches in genetic divergence trees may also be scaled in terms of total number of substitutions observed across the *entire genome*. Either way, both of these scalings represent an amount of sequence change that has occurred.

One thing to watch out for in genetic divergence phylogenies are particularly long branches, often at the tips. As discussed in the previous section, the length of a branch leading to a tip is scaled according to the number of mutations that are observed within that sample's sequence, but not in other samples in the tree. If the numbers of mutations that are unique to a sample are very large, it is often an indicator of a sequence quality issue (e.g., a sequencing error, or a bioinformatics issue with genome assembly or multiple sequence alignment).

2.5.3 Temporally Resolved Phylogenetic Trees

The phylogenetic trees discussed thus far describe an inheritance process for the mutations observed in an alignment of sequences; these are genetic divergence phylogenetic trees. However, there is a second type of tree that you will commonly encounter in genomic epidemiology: **temporally resolved phylogenetic trees**, also referred to as **time trees**.

Time trees make use of the molecular clock that we described earlier in this chapter to translate the amount of genetic divergence observed along branches of the tree into an estimate of the amount of calendar time that likely passed. This translation results in a tree where branch lengths are measured in *absolute time* rather than in genetic divergence. A few other shifts will occur in a time tree as well. Firstly, the position of the tips of the tree (those viruses that you have actually sampled and sequenced) is fixed at the date of sample collection, rather than representing how diverged a sample is from the root sequence. Secondly, the position of internal nodes in the tree is representative of when we think that ancestral node likely existed. The estimated date of an internal node in the tree comes from observing the sequence diversity of the descendents, and asking, "How far back in time would you need to go in order to find the common ancestor of these descendants?" This quantity is referred to as the **time to the most recent common ancestor (TMRCA)**. For trees scaled in absolute time, it is not uncommon to indicate the range of uncertainty associated with the inferred date by providing a 95% confidence interval around the date. This may be given as a numerical range, or shown visually with a rectangle or a violin plot to show the full probability density.

Time trees have several useful properties that support epidemiological inference. Exploring trees along the dimension of time makes the tree more useful from a descriptive epidemiological perspective. For example, when we can move between genetic divergence trees and time trees, we can easily compare how the genomic diversity between samples aligns with the number of serial intervals separating sample collection dates. Furthermore, when we join information about sequenced samples (e.g., geographic area where a case resides) with information on when the common ancestors of those sequences likely circulated, we can begin to reconstruct when spatial movements of a pathogen occurred. This latter procedure is commonly referred to as **phylogeography**.

To discern whether you are looking at a time tree or a genetic divergence tree, look at the x-axis label or scale bar of your tree. If branch lengths are scaled in terms of absolute time (calendar time), then you have a time tree. Alternatively, if branch lengths are given in substitutions, or substitutions per site, then you are looking at a genetic divergence tree.

2.6 THE TRANSMISSION TREE DOES NOT EQUATE THE PHYLOGENETIC TREE

Any branching process can be represented by tree-like graphs. The majority of this handbook discusses phylogenetic trees, which are hypotheses of common descent for genetic sequences. In such trees, sequenced cases at the tips of the tree are connected to inferred common ancestors, which may not have been directly observed. In contrast, within epidemiology we often think about **transmission trees**, in which nodes represent known cases, and edges connecting nodes represent the directionality of who-infected-whom. While it may seem like there are times when a network of who-infected-whom matches the phylogenetic tree, these two types of tree-like graphs are fundamentally different and should not be equated in practice.

What makes a phylogenetic tree different from a transmission tree? First and foremost, genetic mutations are not markers that a transmission event has occurred. Mutations occur spontaneously as the virus replicates, and this process is independent of whether the infection is passed on to another individual. While within-host diversity is sampled and passed on when a transmission event occurs, we do not always observe this variation at the consensus sequence level. As such, mutations in consensus genome sequences by no means occur with every single transmission event. Indeed, it is common for cases with direct epidemiologic linkage to have identical genome sequences.

Secondly, phylogenetic trees are inferred from sequences that are sampled from infected individuals at the time that a specimen is collected, not at the time that a transmission event occurs. This means that an observed sequence tells us something about that individual's infection at the time that their specimen was collected, but not necessarily their infection at the time of a transmission event. Given that an individual's within-host pathogen diversity is accumulating and changing over the entire course of their infection, the sequence that summarises diversity at the time of sample collection (the observable sequence) may not be identical to the sequence that summarises within-host diversity at the time of transmission.

This issue is important because the *sequence* that is ancestral in a phylogenetic tree may not actually represent the *infection* that is ancestral. As a toy example, imagine a transmission event in which Person A infects Person B with some virus. Let's say that Person A and Person B attended a party together just before Person A became symptomatic. Given this timing, Person A infected Person B early on in the course of their infection. However, Person A felt terrible, and didn't seek testing until well into the course of their infection. In contrast, Person B heard that Person A was sick, and decided to seek testing just a couple of days after the party. In this scenario, despite Person A transmitting their infection to Person B, Person B's diagnostic specimen is collected *before* Person A's specimen is collected. Furthermore, given that Person B's sample was collected early on in their infection, their sequence could be identical to the sequence that we would have observed from Person A had Person A been tested earlier. However, over the multiple days where Person A didn't seek testing, they had further viral diversity accumulating, and by the time they were tested, the sequence characterizing their infection would have been additionally diverged (Figure 2.9).

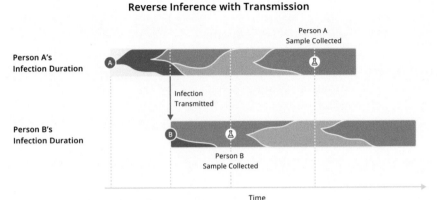

Figure 2.9 An example of how genetic directionality may not represent transmission directionality. In this example, Person A's infection is sampled very late in the course of their infection. Indeed, Person A's sample is collected after they transmitted the infection to Person B, and even after Person B's infection was sampled and sequenced. In this scenario, the sequence representing Person A's infection could look as though it is descended from Person B's infection, even though the true transmission pattern is from Person A to B.

In this case, it's completely possible that while Person A truly did infect Person B, the viral genome sequence collected from Person B truly is ancestral to the sequence collected from Person A. In this way, an accurate portrayal of the pattern of genetic descent does not always show the exact directionality of transmission events.

Although sequencing cannot generally confirm who-infected-whom, there are exceptional cases where this is possible. Chronic viral infections, such as hepatitis C virus or HIV, have a great deal of time to accumulate diversity, making different individual's infections very distinct. With appropriate sequencing methods that can capture this within-patient diversity, in both the infectee and the suspected infector, as well as community controls, it is possible to show that one individual's viral diversity is a subset of another individual's within-host diversity. Sequencing methods for accurately capturing within-host diversity tend to maximise sequencing depth and minimise errors. Public health

applications typically do not sequence in this way for routine surveillance. However, such protocols are more common in academic studies,
and are also used in court cases of criminal transmission.

2.7 WHY IS SEQUENCING BETTER AT DISMISSING LINKS THAN CONFIRMING THEM?

As we just discussed, observing mutations in consensus sequences is not
a perfect marker of transmission events. As discussed at the beginning of
this chapter, the power to use evolutionary analysis to explore epidemiological dynamics comes from the fact that changes to pathogen consensus
genomes occur on *similar* timescales to transmission, but the two processes are not fundamentally joined together such that they occur at exactly the same time. This means that genomic data can provide greater
resolution for understanding relationships between infections, but not
perfect resolution.

The limits of using genomic data to resolve transmission dynamics
are best observed for cases that are epidemiologically linked, or closely-
linked within a single transmission chain. Transmission among these
cases, separated by one or only a few serial intervals, may have occurred
faster than we see mutations arise within consensus sequences, leading
to multiple cases with identical or highly similar consensus sequences.

This limitation imposes an important caveat to keep in mind as you
perform genomic epidemiological inference; sequencing is generally better at dismissing epidemiologic linkage than confirming it.

Let's walk through an example to illustrate this principle. Imagine
you have two cases of a viral respiratory disease from the same household,
with symptom onset dates roughly three days apart. Given the shared
setting that these two people live in, and the timing of their infections, it
seems likely that this pair of cases would represent an instance of household transmission, in which the case with the earlier symptom onset date
likely infected the person with the later symptom onset date. Having genomic sequencing capacity in your public health lab, you take those two
cases' diagnostic specimens and sequence the viral genomes. You look
at the consensus genomes and see that despite these cases coming from
the same household, the sequences are quite divergent from each other.
Given the evolutionary rate of this respiratory virus, you estimate that
you would need roughly six months worth of transmission to accumulate the total number of nucleotide differences observed between those
two sequences. Furthermore, the two cases' sequences are more related

to other genome sequences collected in the surrounding community than they are to each other. You conclude that despite the shared living space and the timing of symptom onset, these two cases are not actually linked, and that it's more probable that they were each independently infected in different settings outside the home. The divergence between their consensus genomes allows you to easily rule out linkage between them.

In contrast, imagine you take that same household pair, and they have identical genome sequences. Their infections are likely related, but you can't tell from the genomic data who infected whom. Moreover, you can't actually tell whether the transmission is even *between* these two cases. For instance, the household pair could both have been infected by a friend at a party they both attended, and they just had slightly different symptom onset rates. With identical genome sequences you can tell that the cases are likely related, but you often won't know exactly *how* related, and it's highly unlikely that you'll be able to rule in a direct transmission event without pulling in additional sources of information.

2.7.1 How Many Mutations are Enough to Rule Linkage Out?

Coming back to ruling linkage out, what is the threshold at which there are "enough" mutations separating sequences that you can consider them to be unlinked? Like many things, the answer is "it depends". And it depends on both the evolutionary rate of your pathogen of interest, and the average length of the serial interval between two cases in a direct infection. Given that both these variables have distributions, it is best to think about sequence divergence and case linkage probabilistically. In the figures below, we show probability distributions for whether cases are directly-linked given that their consensus genome sequences are separated by x mutations. We show these plots for two different (theoretical) viral pathogens with different evolutionary rates and different genome lengths to illustrate how thresholds can change depending on the pathogen and the rate of transmission.

In Figure 2.10, we imagine a viral pathogen with an evolutionary rate and genome length like pandemic H1N1 influenza. We show the probability that two cases are directly-linked given the number of mutations that separate two sequences. Each point in the plot represents a sample from the simulation, and the grey shaded areas represent the probability distributions given those samples. You can see that as the number of mutations separating sequences increases, the distribution of the probability that cases are linked decreases towards zero, but the distributions

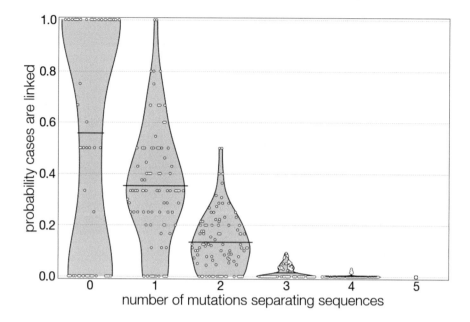

Figure 2.10 Probability distributions for whether cases are directly-linked, given their genomes are separated by a certain number of mutations. The simulation from which these data are drawn recapitulates pandemic H1N1 influenza, with an evolutionary rate of 0.003406 substitutions per site per year, and a genome length of 13,154 nucleotides.

still have overlap. For instance, we still see that even when sequences are separated by two mutations, there is still a roughly 15% probability that the cases are linked. However, once you get to the point where the two sequences are separated by four or five mutations, there is an essentially zero percent probability that the cases are directly-linked.

In Figure 2.11, we show the exact same procedure, however, for a pathogen that is more SARS-CoV-2-like. Compared to pandemic H1N1 influenza, SARS-CoV-2 has a slightly slower evolutionary rate, but a genome that is more than twice as long. Comparing the two plots we can see that, with the second pathogen, the longer genome length (more sites where a mutation could occur) has provided us with a bit more resolution, and we can see that with more mutations separating two samples, the probability of linkage between the two cases drops off more quickly. That said, there continues to be overlap in the probability distributions, and you can see that there is still a non-negligible probability of direct linkage between cases whose sequences are separated by one mutation.

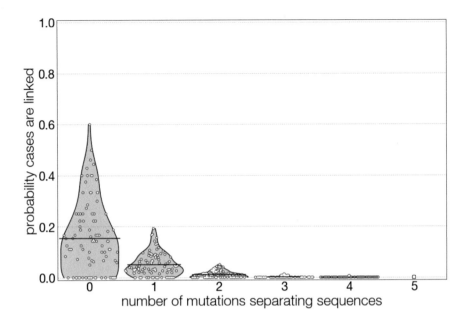

Figure 2.11 The same figure as Figure 2.10, but here the genomic parameters are SARS-CoV-2-like, with an evolutionary rate of 0.0008 substitutions per site per year and a genome length of 29,903 nucleotides.

In conclusion, we hope that these examples provide a clearer understanding of why it is easier to rule linkage out than it is to rule linkage in, and how to think probabilistically about linkage between cases given a certain level of sequence similarity. That said, we recommend that you use this information for shaping your interpretations of data, rather than setting an unvarying threshold by which you consider cases linked or unlinked. Many of the factors that go into shaping these plots are time-dependent. The average length of a serial interval between two linked-cases can rise and fall given changes to the number of individuals who remain susceptible and available to be infected, or the degree of superspreading that occurs, or changes to the intrinsic transmissibility of the pathogen. Furthermore, the evolutionary rates of the pathogens depend on various factors discussed in more depth earlier in this chapter. Taken together, this means that the similarity between the evolutionary timescale and transmission timescale can vary over an outbreak, even for the same pathogen. Thus, these principles should provide guidance and intuition for the data, but not enforce hard and fast thresholds.

CHAPTER 3

Sample Selection

Allison Black

Washington State Department of Health, Seattle, Washington

I N THIS CHAPTER, we describe two major approaches to sample selection for genomic epidemiological studies. We describe when you should use these different sampling strategies, focusing on which types of questions you may be interested in, and how the particular sampling strategy supports the goal of your investigation. This chapter is pertinent for any readers who are actively seeking to implement genomic surveillance programs and integrate genomic epidemiology into their public health investigations. Readers who want to understand why different sampling strategies are useful will also benefit from reading these sections.

3.1 REPRESENTATIVE SAMPLING

In representative sampling, specimens are selected for sequencing such that the pathogen genetic diversity in the sequenced sample set is representative of the pathogen genetic diversity circulating in the broader population. This means that the investigator should be able to detect the same suite of genotypes in their sample as circulates in the broader population, and that the frequencies with which they observe different genotypes in their sequenced dataset should reflect the frequencies with which those genotypes are found in the broader population.

To maintain these attributes and protect the validity of their dataset, the investigator must ensure that they do not accidentally enrich for certain genotypes by preferentially sequencing samples that have a particular diagnostic trait, a particular clinical presentation, or affect a particular demographic group.

Furthermore, to maintain a representative dataset, the investigator must also avoid systematically excluding certain genotypes that are

DOI: 10.1201/9781003409809-3

circulating in the population. Genotype exclusion can occur when you lack sequences from a particular portion of your population that does not mix homogeneously with everyone else in your population. For example, if you have an under-served population that co-mingles with each other, but has limited contact with other groups in your community, then a certain pathogen genotype could circulate primarily within that subpopulation. If that population lacks equitable access to testing, then you may not detect transmission within the subpopulation or sequence the circulating genotype(s). This situation would lead you to miss or underestimate the prevalence of those genotypes.

Generally, we recommend using representative sampling when exploring surveillance questions, such as:

- What is the frequency of this variant in my population?

- How are frequencies of these different variants changing over time?

- What is the spatial distribution of different pathogen genotypes?

- When was this particular pathogen genotype introduced to this population, and how long has it circulated for?

- How much pathogen diversity do we observe in our community, and how does this relate to pathogen diversity in other communities?

3.2 TARGETED SAMPLING

In targeted sampling, we aim to sequence as many samples as possible from a particular population, outbreak, or transmission chain, in order to understand the specific genotypes and disease dynamics associated with that population or setting. Examples of when we use targeted sampling include:

- Investigating populations showing particular clinical presentations of a disease to see whether a specific pathogen genotype appears correlated with an altered disease presentation.

- Exploring whether an outbreak in a localised setting, such as a workplace, school, or medical facility, is the result of transmission occurring within the setting. Alternatively, infections could be acquired in the broader community, and simply detected in the localised setting, for example due to increased screening in the facility.

- Investigating a particular set of cases that report an epidemiologic link to determine whether they are indeed part of the same transmission chain.

- Investigating individuals presenting with a second acute period of illness, to distinguish between a reinfection event and reactivation of latent disease.

3.3 CONTEXTUAL DATA

Targeted sampling focuses on deeply sampling particular settings, transmission chains, or infections that meet case definitions. Yet, to analyse those datasets appropriately, you need to include representatively-sampled genome sequences alongside your targeted samples. Within genomic epidemiology, we often refer to these other representative samples as **contextual data**. They provide a backdrop that describes what is happening more broadly outside of your densely sequenced target population, and can improve analytic inferences. Importantly, they also serve as controls, enabling you to see whether dynamics observed in your targeted population are unique to that population, or whether they are typical of the broader set of sequenced cases.

3.3.1 Contextual Data as a Backdrop

Including contextual data in the analysis of transmission dynamics within a targeted setting enables the public health practitioner to see whether the pathogen genotypes associated with the outbreak circulated in the community before and/or after the outbreak. This contextual information can potentially clarify how an outbreak began, and when it has truly ended. Furthermore, including contextual data in a targeted analysis can help elucidate links between an outbreak in the targeted population and transmission in the broader community. That information can help the epidemiologist to see whether an outbreak amplified transmission and seeded it in the broader population. Additionally, if the bounds of an outbreak aren't truly known, representative samples that appear related to targeted samples may indicate a connection to the outbreak that was not previously known. Finally, including sequences sampled over longer time periods will make estimates of the molecular clock more accurate, and often more precise. When the cases of interest within your targeted analysis occur over a short time window, as in the case of small, localized outbreaks, including contextual sequences sampled over longer periods of time will ensure that your time tree analyses remain accurate.

3.3.2 Contextual Data as Controls

Imagine that you are interested in a cluster of illnesses that have a different disease presentation. You wonder whether a change in the pathogen itself might be responsible for the changed clinical manifestation. To investigate, you decide to sequence pathogen genomes from cases that meet a case definition for the new clinical presentation. In the absence of an association between pathogen genotype and disease presentation, you may find that all of these individuals with similar disease presentations are infected with distinct and diverse pathogen genotypes. However, what if the individuals who meet your case definition are infected with the same or similar genotypes? Does that mean that there is an association between the genotype and disease presentation?

Not necessarily. Only looking at sequences from individuals showing a specific clinical course would be like only looking at the people who got sick after eating at a "poisoned picnic". If every sick case at the picnic ate the potato salad, then you might conclude that the potato salad is to blame. But what if *everyone* ate the potato salad, including your controls who did not get sick after the picnic? Then the potato salad is probably not the culprit. The validity of your study depends on the controls.

Similarly, representatively-sampled contextual sequences act as controls in genomic epidemiological analyses. They allow you to see which pathogen genotypes circulate in individuals *not* presenting with the altered clinical course. Much like the case where everyone at the picnic ate the potato salad, it is completely possible that a particular genotype is dominantly circulating in a community and causing infections with varied clinical presentations. If this is the case, then your targeted samples may all be infected with the same genotype, but your representatively-sampled contextual sequences will also show that same genotype.

CHAPTER 4

Public Health Use Cases of Genomic Epidemiology

Allison Black

Washington State Department of Health, Seattle, Washington

ERE WE DESCRIBE some of the key ways one can use genomic epidemiological analysis in public health investigations. For each of these areas, we provide concrete examples of the types of questions that fall within these topical areas, the fundamental theoretical principles that you will draw upon to investigate those questions, and different sampling schemes and analysis methods for investigating those questions. This chapter is pertinent to most readers, as it describes the public health utility of different genomic epidemiologic analyses. Additionally, readers engaged in building genomic epidemiological capacity or in performing genomic epidemiology themselves will benefit from the descriptions of how to design, analyse, and interpret various genomic epidemiological investigations.

4.1 ASSESSING EPIDEMIOLOGIC LINKAGE BETWEEN CASES

In public health, we often question whether there is a connection between cases presenting with the same infection. Often we'll draw on an assessment of shared exposures, and geographic and temporal data, to determine whether infections are linked. Pathogen genomic data provide another data stream by which we can assess relationships between cases, as we'll discuss here. Some examples of questions that you may ask when using pathogen genomic data to assess epidemiologic linkage include:

DOI: 10.1201/9781003409809-4

- Both of these cases have similar reported exposures. Is that common exposure the likely source of both infections?

- I'm observing multiple detections of disease in the same setting. Is this setting facilitating transmission, or are we simply detecting community-acquired infections in this setting due to enhanced screening or surveillance?

4.1.1 Fundamental Principles to Draw Upon

As a pathogen, let's say a virus, circulates in a population, it infects different people. During those infections, the virus replicates, making errors as it copies its genome for packaging into progeny virions. While different pathogens make errors at different rates, and we may see substitutions in consensus genomes at different rates (see Chapter 2, Section 2.2), this principle means that pathogen genome sequences will accumulate changes over the course of an outbreak. This process also means that cases separated by a minimal amount of transmission will generally have more genetically similar infections, while cases that are separated by large amounts of transmission will typically have more genetically divergent infections.

4.1.2 How Should You Sample?

When you are looking at the pathogen genome sequences of two infections you are not extrapolating from those data to a broader understanding of the outbreak as a whole. Therefore, you can assess possible linkages between cases using sequences collected through either a targeted sampling scheme or a representative sampling scheme. That said, it remains important to include contextual sequence data in your analysis, as described in Chapter 3 and as shown in Figure 4.1.

4.1.3 Tools and Approaches You Can Use to Explore the Question

To investigate questions of linkage you will need a way to compare and summarise the genetic similarity between cases of interest. There are multiple methods for summarizing genetic distances between samples, but we generally recommend using tree structures such as **phylogenetic placements** or **phylogenetic trees** (see Chapter 6 for more details) when assessing sequence similarity between samples.

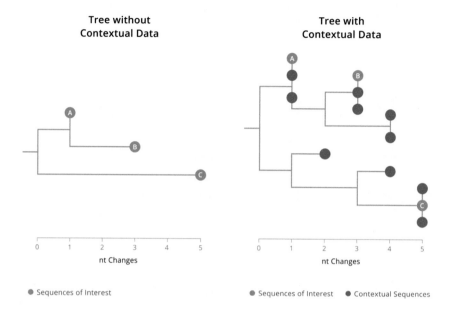

Figure 4.1 A toy example showing the importance of including contextual data when assessing linkage between cases. On the left-hand side is a phylogenetic tree including only three samples of interest. On the right-hand side, we show a phylogenetic tree of the same three samples of interest (orange tips) along with closely-related contextual data (blue tips). The addition of contextual data clarifies the relationships between the samples of interest. In the tree on the left, we might assume that A and B are related cases, since they both share a mutation and are only two nucleotides diverged from each other. However, the addition of contextual data (blue) provides a more nuanced picture. We see that samples of interest A and B are in fact more closely related to contextual sequences than they are to each other. This could mean that A and B are not directly epidemiologically linked, or it might mean that A and B are both part of a larger transmission chain captured by the contextual sequences. Furthermore, contextual data can make it more clear when samples are diverged. While we can see substantial genetic divergence between sample C and samples A and B in the tree on the left, the addition of contextual data makes it more clear that C is unrelated to samples A and B.

While it is technically possible to build a tree structure with just your samples of interest, we caution against doing this, as the addition of other contextual sequences usually makes the relationships between your samples of interest more clear. Furthermore, in the case that your samples of interest *are not* genetically similar to each other, the presence of contextual sequences allows you to see other samples that they *are* closely-related to (Figure 4.1).

Both phylogenetic placements and phylogenetic trees will provide a tree structure for assessing similarity between samples of interest. However, during public health investigations, our questions about the genetic similarity of samples of interest are usually limited in scope and we need answers quickly. When rapid situational awareness is more important than a richer analysis, we recommend performing a phylogenetic placement. If after performing a phylogenetic placement you have further descriptive epidemiological questions about the samples, then we recommend following-up a phylogenetic placement with inferring a phylogenetic tree.

4.1.4   Caveats, Limitations, and Ways Things Go Wrong

As discussed in Chapter 2, genomic data are powerful for ruling linkage out, but less powerful for unequivocally ruling linkage in. Furthermore, except in rare cases, you cannot infer the directionality of transmission from sequence data alone when you have highly genetically similar consensus genomes. This principle is probably clearest when cases have identical genome sequences, since if all of the sequences are the same there's no genomic signal for directionality. However, we caution that directionality is still challenging to infer accurately even when some genomes are slightly diverged. As an example of this issue, in Figure 4.2 we illustrate two different transmission patterns that yield the same phylogenetic tree topology. In the first scenario, we have a person-to-person transmission pattern, where directionality is from Persons 1 to 2 to 3 to 4. In the second scenario, we instead have a superspreading event, in which Person 1 directly infects Persons 2, 3, and 4. Despite having different transmission patterns, the genetic divergence trees are identical. This example shows how challenging it can be to disentangle genetic diversity accrued during person-to-person transmission from within-host diversity that is transmitted when an index case infects multiple secondary cases.

Additionally, in much the same way that contextual data may change your understanding of linkage between cases, lacking sequences from

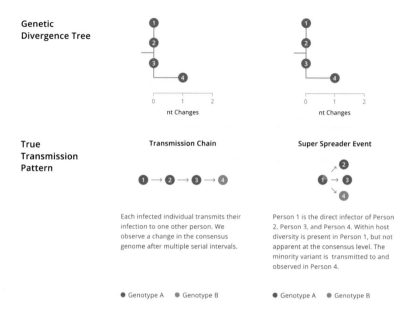

Figure 4.2 Different transmission patterns between Persons 1, 2, 3, and 4 can yield identical genetic divergence trees. On the left, we show a transmission chain in which Person 1 infects 2, who infects 3, who then infects Person 4. Depending on the evolutionary rate of our pathogen, we may not see mutations arise in every single consensus genome. Indeed, here we see that the sequences from Persons 1, 2, and 3 are identical. Thus without knowledge of the true transmission pattern, we would not be able to detect the directionality of transmission between those three individuals. Furthermore, while one might initially think that direction- ality for Person 4 is possible to infer, comparison of the scenarios on the left and right show how directionality can become convoluted by within-host diversity. On the left, Person 4's consensus sequence shows an additional nucleotide change on top of the background genotype of 1, 2, and 3, which was accrued over the process of transmission between individuals. In contrast, on the right that same pattern of diversity is present within Person 1's infection. Then, during a superspreading event in which Person 1 infects Persons 2, 3, and 4 directly, that within-host diversity is captured at the consensus level of the secondary infections. These toy trees are consistent with still more additional true transmis- sion patterns that are not enumerated here. As an exercise, you might want to try sketching additional possibilities out.

infections may also change your interpretation of linkage. When assessing linkage between cases, we recommend keeping your sampling frame in mind. If you're assessing linkage between two samples collected through representative surveillance, then consider what proportion of cases you have sequence data for. If you sequence only a small proportion of cases, then there may be intervening cases along a transmission chain that go unsequenced, and therefore do not appear in a phylogenetic tree. In that case, genetically similar samples could be part of the same transmission chain, but not necessarily directly epidemiologically linked.

Finally, some molecular epidemiological methods suggest using Single Nucleotide Polymorphism (SNP) distances and thresholds for ruling cases into the same outbreak or categorizing them as belonging to different outbreaks. While we understand how thresholds can make decision making easier, we generally discourage using SNP thresholds. This is because using SNP thresholds reduces the granularity of the data; they make a binary assessment (in or out of outbreak) when in fact the data are continuous (0, 1, 2, 3 ... SNPs separating two sequences). While reducing the data down to a single binary outcome does make assessment easier, it does so by hiding nuance and uncertainty, which is critical for making epidemiologic inferences and assessing your confidence in them. Furthermore, making genetic distance into a binary call results in a loss of information. Pathogens accrue mutations at specific evolutionary rates, thus the actual count of differences between two sequences provides a quantitative estimate of how related or unrelated two sequences are.

4.1.5 Relevant Case Studies

- Ruling out a putative index case during a SARS-CoV-2 outbreak.

- Are cases of the same Variant of Concern lineage linked?

4.2 EXPLORING RELATIONSHIPS BETWEEN CASES OF INTEREST AND OTHER SEQUENCED INFECTIONS

In the previous use case, we showed how to think about using pathogen genomic data when investigating a hypothesis of linkage. This use case is similar, but more exploratory in nature, asking instead whether there are previously unrecognised relationships between other sequenced samples and a case (or cases) of interest. If the previous use case was hypothesis-evaluating, you can think of this use case as hypothesis-generating.

Examples of the kinds of questions you might ask that fall into this use case are:

- Is my outbreak related to other outbreaks occurring in other geographic locations?

- Are there any additional cases in the community that might actually be part of this outbreak?

4.2.1 Fundamental Principles to Draw Upon

Exploring relationships between cases of interest and other sequenced infections relies on the same principles as discussed when assessing linkage between cases. The reason that we have separated these concepts within this handbook is not because the principles underlying these concepts are different, but rather because the sequences you are considering in your analysis will likely come from different populations. In the previous use case, public health practitioners are seeking to use genomic epidemiology to understand epidemiologic relationships between individuals within *their own community*, or wherever they have jurisdiction. In contrast, in this use case we describe why and how to explore relationships between your samples of interest and sequenced infections from *other* communities and regions.

4.2.2 How Should You Sample?

Within this broad use case, a question of interest centers around exploring possible links between your own cases (or outbreaks) and cases or outbreaks occurring in other communities. Contextualizing your own transmission in this way brings up a key concept and challenge; if you are looking for linkage with transmission in other contexts, then you want access to sequences that accurately capture the full scope of pathogen diversity circulating in those other contexts. This means that while you can select your own cases of interest in a way that aligns with the question you're investigating, you typically need contextual data from other jurisdictions that have been sampled *representatively*. When sampled representatively, contextual data provide a more accurate summary of the circulating pathogen diversity in those other areas. When available sequence data from other regions accurately summarises those outbreaks, you can have greater confidence that you are accurately capturing the true presence or absence of cross-jurisdictional links.

This concept brings forth a few important considerations. Firstly, the need for representatively sampled contextual data is one of the reasons why we need broad, baseline genomic surveillance programs. In such programs, sequencing is performed at random, without regard to specific clinical presentations, performance on diagnostic assays, or public health questions. Representative sequencing is done partially for the common good; everyone will need contextual data from other communities or populations at some point. This is the main reason why we encourage groups building genomic surveillance systems to perform some degree of representative sequencing beyond their targeted outbreak sequencing. Furthermore, this rationale is also why *annotating* those sequences as representatively sampled surveillance specimens is a critical part of genomic surveillance data management.

4.2.3   Tools and Approaches You Can Use to Explore the Question

To answer these types of questions, we recommend phylogenetic approaches that summarise the genetic relationships between samples on a tree structure. Currently, there are two primary ways to infer a tree structure that includes your cases of interest: phylogenetic placements and phylogenetic trees. While the output of these two types of analyses may look similar, they are performing different processes. Each of these tree structures, as well as tools for performing phylogenetic placements and inferring phylogenetic trees, can be found in Chapter 6.

As an additional layer to building a phylogenetic tree, genomic epidemiologists will frequently investigate cross-community outbreak linkage using **phylogeography**. Phylogeographic analyses take in geographic information about where sequences were sampled and probabilistically reconstruct where ancestral pathogens in the tree likely circulated. More information about phylogeography and implementing phylogeographic analyses is given in Chapter 6.

4.2.4   Caveats, Limitations, and Ways Things Go Wrong

While designing and maintaining representative sampling within your own jurisdiction can be challenging, influencing how sampling occurs externally is close to impossible. Variability in access to resources often leads to variability in which communities have pathogen genomic representation, which can directly impact your own inferences. While resourcing broad genomic surveillance programs at higher levels of jurisdictional authority (e.g., nationally) and clear annotation of representatively-collected data can mitigate this issue, the lack of control that you'll generally have surrounding which contextual data actually exist publicly means that you should exercise some caution in interpreting your findings. This does not mean all is lost! For many investigations, genomic epidemiologists have simply analysed whichever sequences existed or could be generated. And despite such non-ideal sampling, those analyses have still contributed greatly to our understanding of a pathogen's epidemiology. We simply recommend recognising that the data you have access to may be incomplete or biased, and that your interpretations may change upon the addition of more data.

As the amount of publicly-available sequence data increases, you may find yourself shifting from including *all* contextual sequences that you can find to having to choose *which* contextual data to include. This will be true in particular for phylogenetic tree-based analyses, where there is a practical limit on the number of sequences you can include. For guidance on how to choose which contextual data to include, please refer to Section 6.4 within Chapter 6.

4.2.5   Relevant Case Studies

- Identifying, assigning, and investigating a new SARS-CoV-2 lineage in Lithuania

- Evaluating an intake screening program to prevent introduction of SARS-CoV-2 to prisons.

4.3   ESTIMATING THE START AND DURATION OF AN OUTBREAK

Once we have detected that an outbreak is occurring, we often want to know more about it. How did this outbreak begin, how long has it been going on for, and when will we be able to declare it over? Pathogen genomic data can help us to understand outbreaks, especially the timing of when they began, which can be challenging to determine from case surveillance data when rates of asymptomatic infection are high or when most cases are not readily captured by passive surveillance. The types of questions your might find yourself exploring within this use case include:

- Our syndromic surveillance system isn't as sensitive as we would like, and we aren't sure how long we've had circulation of this pathogen in our community. When was this pathogen introduced into our community, and how long has it been circulating for?

- We received reports of some initial cases of a disease, but we weren't able to confirm an outbreak until later. Did this outbreak begin around the time of those initial case reports?

- We detected an outbreak and we believe that we brought it under control, however we continue to have sporadic cases in the facility. Is the outbreak still ongoing?

4.3.1 Fundamental Principles to Draw Upon

Genomic data are useful for estimating the timing of epidemiologic events in a few different ways. Firstly, estimating the timing of disease circulation from genomic data can be more accurate when the disease in question causes large numbers of asymptomatic infections or mild cases that do not seek treatment. In these scenarios, surveillance systems may only start to record cases once a sufficient number of infections have occurred to produce a subset of more severe cases that seek care or diagnostic testing. In contrast, even when an infection is asymptomatic or mild, viral replication will occur, leading to mutations that may be carried forward during transmission. In this way, a record of the infection can be left in the pathogen genome even when the infection does not rise to a sufficiently symptomatic level to be a recorded case. Furthermore, since genomic data can enable you to resolve distinct but concurrent outbreaks, temporally resolved phylogenetic trees can provide *cluster-specific* estimates of timing. This feature is particularly useful once you have ongoing transmission, and the emergence of distinct clusters (such as the emergence of a new variant) may not be apparent within an epidemiologic curve.

In Chapter 2, we introduced molecular clocks, which represent the average rate at which a particular pathogen evolves. When we know on average the rate at which genetic diversity accumulates, we can take an observed amount of genetic diversity and ask: "How long would it have taken to accumulate this much diversity?" When we are looking at the genetic divergence between sequences sampled from the same outbreak or epidemic, calculating how much time was needed to generate that diversity provides an estimate of when that outbreak or epidemic likely started.

We generally investigate this type of question using a temporally resolved phylogenetic tree, and we frame the question as "What is the time to the most recent common ancestor of these sequences?" Or, "When did the most recent common ancestor of these sequences likely circulate?" When sequences are genetically similar, little evolutionary time has passed, and the ancestor from which those sequences are descended will be more recent. When sequences are very diverged, a large amount of evolutionary time has passed, and the ancestor of the sequences of interest will have existed further back in time.

Every internal node within a temporally resolved phylogenetic tree represents an ancestor and will have an attached date. So, which internal node should you look at? Typically, the ancestral sequence that you are looking for is the youngest, or least basal, internal node from which all sequences of interest descend. For example, if you are interested in estimating when an outbreak in a skilled nursing facility began, then the internal node that you would look for is the youngest node in the tree from which all skilled nursing facility-associated samples descend. Or if you are interested in when Zika virus likely arrived in the Americas, then you would look for the ancestral node from which all Zika virus genomes collected across all countries in the Americas descend.

4.3.2   How Should You Sample?

In order to estimate the molecular clock accurately, you will need genome sequences collected over longer time spans. The reason for this is fairly simple; it is very hard to accurately estimate the slope of a line when the only data points you have come from a single cross-sectional sample. In contrast, data points collected over time allow you to see much more clearly how genetic diversity and time correlate, thereby providing more consistent and accurate estimates of the evolutionary rate. Ideally, samples used in estimating the molecular clock should be sampled representatively. These serially sampled sequences can either be sequences you generated, or they can be publicly available contextual sequences.

Once you have your estimate of the evolutionary rate, the samples of interest (whose ancestor you would like to date) can be sampled either representatively or in a targeted fashion, in accordance to your question of interest. If you are interested in when a localised outbreak began, then you may want to intensively sample and sequence infections that occur within that particular facility or setting. However, when the event that you would like to date has yielded many sequenced samples, it is best to sample the sequences representatively. For example, if I wanted to estimate when SARS-CoV-2 was first introduced to California, I should select a representative sample of sequences from the entire state of California, and estimate when their ancestor likely circulated. While in theory I could perform targeted-sampling of the entire state of California and sequence every positive case, in reality that would be completely infeasible. Thus, to accurately estimate how much time was necessary for Californian viral diversity to accrue, I must have a representative sample of Californian SARS-CoV-2 diversity. If my sample is *not* representative, then the date that I infer will represent when the ancestor of the sample that I have circulated.

4.3.3 Tools and Approaches You Can Use to Explore the Question

Using the molecular clock to translate the branch lengths of trees from genetic divergence (evolutionary time) to calendar time is a fairly common method within genomic epidemiology. As such, there are various phylogenetic inference tools that you can use for building time trees. Given the rapid turnaround times that we typically desire in public health, Nextstrain analyses are currently the most common way of inferring time trees within public health contexts (see Chapter 6). Currently, phylogenetic placements do not allow you to make time trees.

4.3.4 Caveats, Limitations, and Ways Things Go Wrong

As discussed above, it is important that you always remember that you are inferring the timing of ancestors of your *sampled* sequences. Any pathogen genetic diversity that circulated in a population, but is not captured in your dataset, will not be included in your inferential procedure. This is why representative sampling of large populations is important for accurately estimating when transmission started.

As a concrete example, imagine that you want to estimate when SARS-CoV-2 first emerged into the world. But also imagine we're back in October 2021, and the Delta lineage of SARS-CoV-2 has completely taken over. All other variants that previously circulated have largely gone extinct. All contemporaneously sampled viruses are Delta lineage.

When building your dataset for the analysis of when SARS-CoV-2 first emerged, you download a representative sample of all publicly-available SARS-CoV-2 genome sequences from the past two months. You see that the estimated age of the root of that tree seems to be in spring of 2021, but you know that SARS-CoV-2 was circulating for more than a year before that. What went wrong?

All of the sequences in your theoretical analysis are Delta-lineage descendents. In your large analysis, you have estimated when the ancestor of all of that *Delta-lineage diversity* likely circulated, but that ancestor isn't the same as the ancestor of *all SARS-CoV-2 diversity that ever existed*. In order to estimate when SARS-CoV-2 originally emerged, you would need to ensure that you include sequences that represent the viral diversity that circulated before Delta took over.

An additional point to keep in mind is that molecular clocks vary depending on multiple factors (discussed in Chapter 2). Different datasets might give you slightly different estimates of the rate of molecular evolution, and subsequently slightly different estimates of when ancestral viruses circulated. This variability is an inherent part of these analyses, and so we recommend always providing confidence intervals around your date estimates. When you see this variability, do not panic! If you estimate a faster evolutionary rate in a particular dataset it is unlikely that the pathogen is now suddenly evolving more quickly. The more likely explanation is that you have dense genomic sampling over a short time frame, meaning that purifying selection has not purged deleterious mutations from the pathogen population. Since there hasn't been enough time for purifying selection to act, you'll see more diversity and estimate a faster evolutionary rate. This scenario occurred during the epidemic of Ebola Virus Disease in West Africa in 2013–2016 [12].

4.3.5 Relevant Case Studies

- Estimating when the Zika Virus outbreak began in Colombia.

4.4 ASSESSING HOW DEMOGRAPHIC, EXPOSURE, AND OTHER EPIDEMIOLOGICAL DATA RELATE TO A GENOMICALLY DEFINED OUTBREAK

Much of the richness of genomic epidemiological analysis comes not simply from the analysis of genomic data alone, but rather from the integration, and joint inference, of genomic and epidemiologic data together. This use case therefore describes the situation in which you seek to bring genomic and epidemiologic data together to understand how exposures or demographic information relate to genomically defined relationships between samples. Examples of the kinds of questions that you might ask within this use case are:

- Are different lineages of this pathogen circulating in younger people versus older people?

- Are different lineages of this pathogen circulating in vaccinated individuals versus unvaccinated individuals?

- Does outbreak size or transmission intensity appear to correlate with circulation in particular populations of individuals?

4.4.1 Fundamental Principles to Draw Upon

Rather than relying on any specific principle within genomic epidemiology, this use case presents how one can bring inferences from genomic epidemiology together with surveillance or other epidemiological data. In doing these analyses, the user is looking at genomic relationships between samples, and overlaying those relationships with additional information about where cases lived or worked, their demographic information, possible exposure settings etc. This allows the epidemiologist to qualitatively assess potential relationships between an exposure or demographic variable and clustering patterns in the tree.

4.4.2 How Should You Sample?

You can bring together genomic and surveillance data across any kind of sample set, although you may find different utility in adding surveillance data to outbreaks sampled in a targeted way as compared to representatively-sampled surveillance samples. For example, if you are seeing multiple outbreaks across various skilled nursing facilities (SNFs) in your jurisdiction, you may wish to conduct dense, targeted sampling of cases among employees and residents of those various facilities. Then, you may wish to see how the facility that cases are associated with interacts with the genomic relationships you observe. Do individuals from a single SNF tend to cluster together in clades that are genetically diverged from cases in the other SNFs? Or do all the cases cluster together within a single clade, and adding data about which SNF a case came from shows that cases from all of the SNFs are highly intermingled? In this targeted-sampling example, you are interested in how an exposure of interest (SNF) is associated with close genetic relationships between infections.

In contrast, when joining epidemiologic data and representatively-sampled genomic data, you may be less interested in exposure-outcome associations, and more interested in how well you are capturing a representative cross-section of your population. Adding demographic data to a tree can help you qualitatively see trends such as whether you're capturing pathogen genetic diversity sampled only from urban centers, or from rural areas as well, and whether you're capturing cases from different age groups and racial, ethnic, or national-origin groups. Doing this kind of procedure allows you to investigate whether you are likely capturing the full breadth of circulating viral diversity, or whether there is a population that appears to systematically lack genomic surveillance representation.

4.4.3   Tools and Approaches You Can Use to Explore the Question

One of the most common approaches for bringing together surveillance data with genomic data in public health applications is to use the "metadata overlay" feature in Nextstrain. This feature allows the user to colour the tips of a Nextstrain tree visualised in Nextstrain Auspice (https://auspice.us/) according to additional variables specified in an external spreadsheet. One of the reasons why public health practitioners use this particular workflow is because it provides a way to join genomic data objects (the tree) with epidemiologic data, which often contains personally identifiable information (PII) or personal health information (PHI) that epidemiologists must keep secure. This need for storing PII/PHI on secured computational infrastructure usually precludes it from being incorporated into the tree directly, since bioinformaticians usually infer trees on scientific-computing infrastructure that is not authorised to store PII. In the case of the Nextstrain metadata overlay, the data table containing the surveillance data remains "client-side", that is, the information never leaves the secure computer. Explicit instructions about how to format and use the Nextstrain metadata overlay are described in Chapter 6.

4.4.4   Caveats, Limitations, and Ways Things go Wrong

What would an epidemiologic handbook be without at least one mention that correlation does not imply causation? In the case where you bring genomic and surveillance data together and look at how the exposure data relate to the phylogenetic patterns, you are in essence looking *qualitatively* for patterns of correlation between the surveillance data and the genomic data. To say that this is qualitative and correlative is not to undermine its utility; indeed, when brought together these data sources typically work synergistically to provide rich information regarding what transmission dynamics may be at play. However, as with any qualitative, observational analysis, the observed dynamics could be subject to confounding. As such, we recommend using this tool to derive quick situational awareness, and suggest that epidemiologists follow-up with more rigorous studies if the relationships observed warrant deeper investigation.

4.4.5   Relevant case studies

- Evaluating an intake screening program to prevent introductions of SARS-CoV-2 to prisons.

CHAPTER 5

Case Studies in Applying Genomic Epidemiology

Allison Black

Washington State Department of Health, Seattle, Washington

Gytis Dudas

Vilnius University, Vilnius, Lithuania

T HE BEST WAY to really get to know theory and principles is by applying them. In this chapter, we provide case studies exemplifying how the broad thematic areas of genomic epidemiological analysis described in the previous chapter can manifest in practice. These case studies illustrate step-by-step how different questions were investigated with a genomic epidemiological approach. Rather than providing the kinds of neat and cohesive narratives we frequently see in peer-reviewed literature, we have tried to show how hands-on investigations really proceed, including evaluating competing hypotheses and describing sources of uncertainty in our analyses. This chapter is pertinent to readers who will be directly involved in analysing and interpreting genomic epidemiological studies, or who want to see examples of genomic epidemiology in practice.

5.1 ARE CASES OF THE SAME VARIANT OF CONCERN LINEAGE LINKED?

At the beginning of 2021, public health authorities within the United States were concerned about the transmission of the B.1.1.7 lineage

("Alpha" within the WHO nomenclature system), which appeared to be more transmissible than previous strains of SARS-CoV-2. Already the dominantly-circulating lineage within the United Kingdom, public health agencies within the United States were interested in whether B.1.1.7 had already arrived in the United States, and if so, how it was distributed and at what frequency. Genomic surveillance activities were established and intensified to investigate these questions. Beyond estimating frequency, some public health agencies also monitored for individual cases infected with "Variant of Concern" (VOC) lineages, which they might prioritise for more aggressive contact tracing and control efforts to limit the establishment and growth of VOC lineages within the jurisdiction.

County A is a predominantly rural county, in which diagnostic testing and genomic surveillance are primarily handled by the local public health laboratory. The county started an in-house SARS-CoV-2 sequencing program at the beginning of 2021, and had regular sequencing and lineage assignment protocols in place during the spring of 2021, when they detected their first two cases of B.1.1.7 lineage viruses among two residents of the county. The two cases were detected and sampled during the same week, a timeline that was consistent with one case infecting the other, resulting in an epidemiologically linked pair. Beyond the timing of the two infections, epidemiological-linkage between the cases seemed possible since both cases were assigned the same lineage (Pango lineage B.1.1.7), and no other B.1.1.7 cases had been detected in County A up to this point. However, while Pango lineage assignments can provide a useful summary of different genetic lineages, most Pango lineages have genetic diversity within the lineage. Especially for lineages whose frequency grows significantly, such as B.1.1.7, there may be many different transmission chains of B.1.1.7 viruses circulating within different geographic areas. In such cases, phylogenetic analyses can provide higher resolution for refining relationships between cases.

In response to these detections, County A undertook a rigorous analysis of the whole genome sequences from these two cases. Firstly, since these were the first B.1.1.7 cases detected in the county, County A wanted to verify the accuracy of the lineage calls. To do so, they looked at the .bam files, which show the actual sequencing read data mapped to a reference genome. To confirm the quality of the B.1.1.7 lineage defining Single Nucleotide Polymorphism (SNP) calls, they looked at the total number of reads that covered each lineage-defining site, and looked for

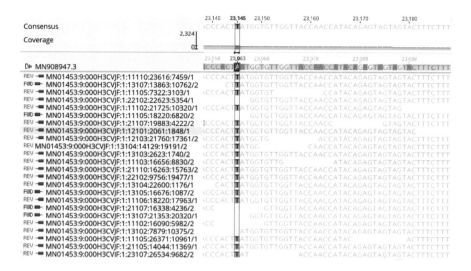

Figure 5.1 A zoomed in picture of site 23063 in the nucleotide sequence, where the reference sequence has an A and the sequencing reads show that this infection has a T at this site. The A23063T mutation in the nucleotide sequence corresponds to an N501Y substitution in Spike protein, which is one of multiple lineage-defining mutations for B.1.1.7. At this site there are 390 distinct, high quality sequencing reads that support this call. Furthermore, both forward (FWD) and reverse (REV) reads detect this nucleotide, further demonstrating that this call is real.

the particular SNP call in each of the reads (Figure 5.1). This process was repeated for all B.1.1.7 defining sites.

Next, County A took the two confirmed B.1.1.7 lineage viruses and imported the sequences into Nextclade (https://clades.nextstrain.org/) in order to assess their quality. The sequences showed minimal numbers of Ns, no mixed sites, and a reasonable number of mutations as compared to the Wuhan-Hu-1 reference genome (Figure 5.2). This indicates that the sequences are high quality, and appropriate for analysis. Notably, the sequences do have frameshift mutations in them, which is why the Nextclade F metric is red (Figure 5.2). While sometimes real, these frameshifts are often a bioinformatic artifact from the consensus genome assembly pipeline. While the author of a genome sequence will typically have to fix or document a frameshift for public repositories to accept the sequence, one can still use them in phylogenetic pipelines that "strip" these frameshifts away.

Figure 5.2 The two B.1.1.7 lineage viruses as visualised in Nextclade (sequence names are intentionally masked). Notably, all quality metrics show that the sequences are of high quality, except for the "F" metric, which indicates that a frameshift mutation has been detected. The sequence view shows that the two sequences share many SNPs, but also have some unique SNPs that are not found in the other sequence.

Looking at the two sequences together in the Nextclade alignment viewer, we can see that both samples share some SNPs, as indicated by the coloured bars that vertically align (Figure 5.2). This makes sense since we know that both samples are B.1.1.7 lineage viruses. However, we can also see that there are differences between the two sequences as well (Figure 5.2). Each sample has numerous additional SNPs that are unique to the sample. This indicates that these samples are likely not directly related. The next paragraph will discuss that logic in more depth, and show how we can see that genetic divergence on the phylogenetic placement available in Nextclade.

In Nextclade, we can take our two sequences and "graft" them onto a pre-inferred Nextstrain phylogeny in a process that is termed "phylogenetic placement". The sequences are placed onto the tree according to the patterns of substitutions that the tree summarises, and that your sequences have. The sequences of interest are placed onto the tree at the point where most of the SNPs in your genome sequence have also been observed in the tree. Then, any mutations that are unique to your sample, and not yet detected in the background tree, are shown as branch length leading from the tree to the sample of interest. In Figure 5.3A, we can see that both sequences group in the Alpha lineage portion of the tree, consistent with their designation as B.1.1.7 lineage viruses. When we zoom in to the Alpha clade of the tree, we can see that despite both viruses receiving a B.1.1.7 assignment, they are quite diverged (Figure 5.3B). Indeed, these two B.1.1.7 sequences are separated by 13 SNPs (Figures 5.2 and 5.3B). This is much more genetic divergence than we would expect to see if the sequences were epidemiologically linked.

Figure 5.3 (A) Nextclade phylogenetic placement of the two B.1.1.7 sequences onto a background Nextstrain tree. Nextclade places both of these sequences on the tree within the Alpha clade, consistent with the Pango lineage assignments and our visual assessment of the sequencing reads. (B) A zoomed in view of the two sequences within the Alpha lineage of the tree shows that they are placed onto different parts of the tree within the Alpha clade. You can count the number of nucleotides changes separating the two viruses by "walking the path" of the highlighted branches between the two samples, counting up the number of nucleotide changes observed on each segment of the path as you go. An important note is that the vertical axis has no meaning, thus you are only summing up the number of SNPs observed along the horizontal segments of the path (the branches).

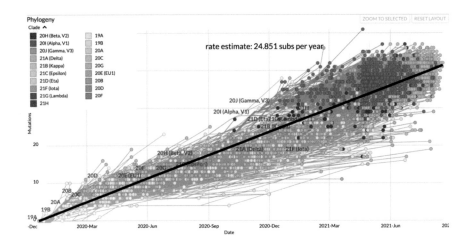

Figure 5.4 The clock view from the global SARS-CoV-2 Nextstrain build maintained by the Nextstrain team (https://nextstrain.org/sars-cov-2/). Each circle in this plot represents a sampled virus, the same as a tip in the tree, and the x and y positions in the plot are determined by the date the virus was sampled and the number of SNPs that it has compared to the Wuhan-Hu-1 reference genome respectively. The evolutionary rate estimate comes from the slope of a least-squares line fit through the data points.

For an in-depth discussion of different thresholds of divergence, please see Chapter 2 (Section 2.7). However, we can also do a back-of-the-envelope calculation to provide some context about what a 13 SNP difference means. Firstly, we should note that the average evolutionary rate of SARS-CoV-2 is roughly 24.5 substitutions across the entire genome per year (see Figure 5.4). This means that if we were to randomly sample two sequences that were sampled a year apart from each other, we would expect them on average to be 24.5 SNPs different from each other. If you were looking at the actual divergence between two sequences, the SNP difference counts would be whole numbers, but this rate is the average value if you were to repeat that sampling procedure many times over. If we take our two samples from County A that are separated by 13 SNPs and think about what this distance means in light of the evolutionary rate of the virus, we can say that roughly six months worth of transmission separates these two infections.

What does this mean from a public health standpoint? County A can be confident that these two cases are not linked – the genome sequences are too diverged for that to be likely. Thus, these two cases of B.1.1.7 likely represent independent introductions of B.1.1.7 into the county. This means that these two cases do not represent a transmission event of B.1.1.7 within County A. Rather, these two cases were likely separately infected somewhere else outside the county, and then returned home where they were tested for COVID-19. At this point, case interviews could be helpful for discerning whether these cases recently traveled, or what exposure event may have led them to contract a lineage that had not previously been detected in County A. Furthermore, since these two cases do not represent sustained transmission within the county, aggressive contact tracing efforts might be warranted in order to prevent establishment of B.1.1.7 circulation in the county.

5.2 EVALUATING AN INTAKE SCREENING PROGRAM TO PREVENT INTRODUCTION OF SARS-COV-2 TO PRISONS

County C has a large jail housing a considerable number of pre-trial detainees. Given the challenge of controlling SARS-CoV-2 outbreaks in congregate settings, including prisons, public health officials in County C implemented a screening system for COVID-19. The jail had experienced previous outbreaks, and the intent of this screening program was to prevent new introductions of SARS-CoV-2 into the jail, which could seed further outbreaks. As part of the screening program, newly incarcerated individuals were tested for COVID-19 if they consented, and underwent a two week quarantine period prior to being moved into the standard housing units.

The jail recorded cases across multiple residential pods. Given the duration of time over which cases were observed, and the fact that multiple residential pods appeared to be affected, epidemiologists were interested in what processes were contributing to COVID-19 cases within the jail. Was the intake quarantine program ineffective, in which case newly-admitted persons were introducing SARS-CoV-2 into the jail? Or had a previous outbreak within the jail never fully ended?

Given the ability of pathogen genomic sequence data to differentiate between related and unrelated cases, investigators sequenced samples from cases who tested positive during their intake screening quarantine period, and from cases that were detected within the residential pods of the jail. Five sequences were collected from individuals who had tested positive during intake quarantine, and 21 sequences were collected from individuals who tested positive while residing in the jail. All of the sequences were assigned the same Pango lineage, and thus the lineage information did not provide sufficient information to discern relationships between the samples. Therefore, investigators took these sequences and performed a phylogenetic analysis of jail sequences alongside contextual sequence data that had been collected from County C and other locales through representative surveillance sampling.

The genetic divergence tree in Figure 5.5 shows the clade grouping 26 sequences sampled from incarcerated individuals (jail sequences are in yellow, contextual data are in grey). Looking at that clade we can see a few important findings. Firstly, some of the jail sequences are genetically diverged, and are more closely related to contextual sequences sampled from the broader community than they are to other sequences collected from incarcerated individuals. Secondly, there is one clade in which samples from the jail have identical consensus genome sequences and sequences that are very closely related to each other. Knowing that genetically similar infections are more likely to be linked to each other, and that genetically dissimilar infections are likely unrelated, we can surmise from this tree that some of these cases likely constitute an outbreak, while other cases among incarcerated individuals represent community-acquired infections that are not part of the outbreak.

Knowing that not all of these cases are genetically related is the first step in evaluating the intake screening program. However, this information alone is not sufficient. What we really want to see is *which* of these 26 sequences come from cases detected during intake screening, and which cases come from incarcerated individuals living in the residential pods. To combine the genomic picture with this epidemiologic data, we can make use of the functionality in Nextstrain that allows us to overlay additional surveillance data onto a tree. In Figure 5.6, we've overlaid the phylogenetic tree with a .tsv file that maps "case status" (either inmate or intake screening) to the "strain name" of the sequence in the Nextstrain tree. This allows us to then colour the tree according to this new field. In Figure 5.6, we can see that sequences from cases found during intake screening are coloured in yellow, while cases detected within the jail are coloured in blue.

This picture helps us understand the following. Firstly, cases detected through the intake screening program **are not** genetically related to cases that were detected within the jail's residential pods. This means that the observed transmission within the jail's residential units **was not**

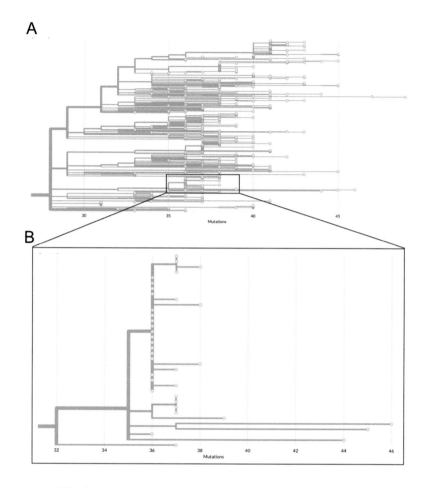

Figure 5.5 Phylogenetic tree of jail sequences and representatively-sampled contextual sequences. (A) The top panel of this figure shows the clade within which all jail sequences group with other contextual sequences. Contextual sequences are shown in grey while jail sequences are coloured yellow. In this view, you can see the broad distribution of some jail sequences across the entire clade, as well as one cluster of jail sequences that appear closely genetically related. (B) A zoomed-in view of the clade which clusters many jail sequences together. Many of the jail sequences within this cluster have identical genome sequences, and therefore appear stacked vertically along the root node of the outbreak clade. We see other jail sequences within this cluster that appear to have one or two additional nucleotide mutations.

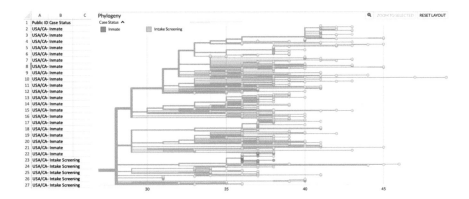

Figure 5.6 Metadata overlay onto Nextstrain phylogenetic tree differentiates sequences from incarcerated individuals (inmates) and from intake
screening. On the left hand side is a screenshot of the metadata file
formatted such that it can be dragged and dropped onto the tree. On
the right, we see the same cluster as shown in Figure 5.5, but now jail
sequences are coloured according to whether they were sampled from incarcerated individuals residing in pods (blue) or from individuals under-
going intake screening (yellow). Contextual sequences remain coloured
in grey.

initiated by any of the sequenced positive cases detected during intake
screening. Secondly, intake cases are genetically similar to sequences sampled representatively from surrounding communities. This result supports a scenario in which newly admitted individuals were infected prior
to their intake, with those infections discovered during the quarantine period. Thirdly, sequenced cases from intake screening are terminal tips in
the tree. By that, we mean that there are no sequences in the tree that
descend from infections detected during intake screening. This means
that the quarantine period halted onward transmission of SARS-CoV-2.

In contrast, within the clade grouping all sequences sampled from incarcerated individuals living in the residential pods, we see that 12 cases
are infected with an identical genotype that is basal to other related
genotypes detected among the incarcerated individuals residing in pods
(Figure 5.7). Additional samples from incarcerated individuals show
additional substitutions accrued on top of the basal genotype. While
some of this diversity could represent the transmission of within-host

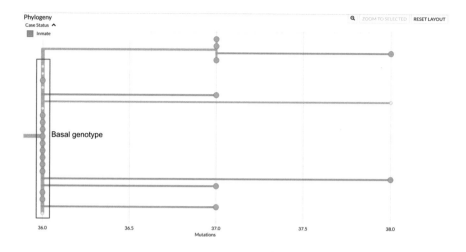

Figure 5.7 Phylogenetic tree showing the outbreak within the residential pods of the jail. Twelve sequences sampled from incarcerated individuals in the residential pods show identical genome sequences, and are the basal genotype of this clade. There are five contextual sequences that also have this identical genotype. Further investigation of which cases these contextual sequences were sampled from may provide some information regarding how this outbreak was introduced to the jail. Three sequences have an additional unique nucleotide mutation, and one sequence has an additional two nucleotide mutations that are unique to that sample. At the top of this clade we see an additional clade that groups four jail samples that all share one additional nucleotide mutation on top of the basal sequence. This pattern of unique and shared diversity, in conjunction with the timeframe over which these samples were collected, suggests ongoing circulation of this clade within the jail over time.

diversity during a super-spreading event, it is also consistent with substitutions accruing over the course of multiple transmission events. The latter scenario, that the accrued genetic diversity is the result of ongoing circulation within the jail, is more likely given that the cases among incarcerated individuals were detected over multiple months.

In terms of public health policy, these findings indicate that the intake screening program served its intended effect, namely, preventing new introductions of SARS-CoV-2 into the jail. While it is unclear how the outbreak was introduced into the prison, most cases within the incarcerated population appear to be related to ongoing transmission within the jail, and thus interventions should focus on controlling transmission within the jail and its residential pods.

5.3 IDENTIFYING, ASSIGNING, AND INVESTIGATING A NEW SARS-COV-2 LINEAGE IN LITHUANIA

Many countries started or expanded their existing sequencing programs following the announcement of SARS-CoV-2 lineage B.1.1.7 (Pango nomenclature) aka Alpha (WHO nomenclature) lineage in late December 2020. At the time, the pandemic had been in its first year of circulation in humans and numerous lineages had acquired sufficient mutations to be easily told apart by sequencing, some of the mutations seemingly being selected by high seroprevalence rates in some countries. One striking feature of this diversity had been the same mutations arising independently in multiple lineages, suggesting that similar solutions to common evolutionary pressures were arising in the virus population. In this case study, we will examine lineage B.1.620, a globally rare lineage with a striking constellation of mutations and deletions most of which had been seen in individual Variants of Concern (VOCs) at the time, but not in combination and without evidence of recombination. We will examine the circumstances under which lineage B.1.620 was discovered, how its origins were determined despite limited data, and how to navigate the landscape of competing hypotheses using phylogenetics.

The Lithuanian SARS-CoV-2 sequencing program, organised by the National Public Health Laboratory and started in February 2021, employed the sequencing capacity of domestic hospitals and universities, as well as the resources of the European Centre for Disease Control (ECDC) reference laboratory. At the time, the program's maximum capacity enabled sequencing of roughly 700 SARS-CoV-2 genomes per week. With this degree of sequencing performed on representatively-sampled specimens, and given the SARS-CoV-2 incidence at the time, this level of sequencing should have been sufficient to detect lineages that represented 0.36% of the cases (0.0036 is the upper 95% binomial confidence interval using Jeffrey's method at 0 detections out of 700 samples). During the study period described here, the predominant lineages in Lithuania were B.1.1.7 (designated as Alpha in WHO VOC nomenclature) and B.1.177.60, a Lithuanian sublineage of B.1.177 which itself is of Spanish origin and came to dominate in much of Europe in late 2020 due to holiday-goers visiting Spain in summer of 2020 [2].

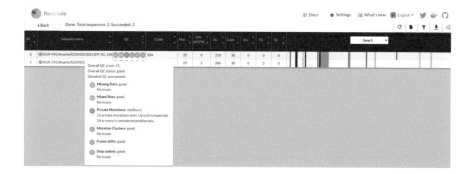

Figure 5.8 Spike protein changes and Nextclade warnings of the two mystery genomes. Because the reference tree has been updated since these genomes were described, private mutation warnings are not as extreme as they were at the time of discovery.

Sequencing results arriving during the week of April 5, 2021 included the typical mixture of predominant B.1.1.7 and B.1.177.60 lineages, with two genomes designated as B.1.177.57 by Pango, but which unusually included the amino acid change E484K in the Spike protein of the virus. These two unusual genomes were also flagged by Nextclade as having too many private mutations (i.e., mutations that are not shared with any others in the tree) (Figure 5.8) while their placement on a bigger phylogenetic tree using strict similarity implied that these two sequences could not possibly be B.1.177.57 because they fell in a completely different part of the tree than even basal lineage B.1.177. This mismatch highlighted the differences between methods of classification used by Pangolin and by Nextclade. At the time using a machine learning-based approach, Pangolin could not cope well with out-of-sample prediction (i.e., identify lineages it hasn't seen before) and its underlying decision tree architecture made it prone to classify distinct lineages as the same based on independent, repeat mutations. Nextclade, on the other hand, attempts to place each SARS-CoV-2 genome query on a reference tree by identifying nodes whose sequence is most similar to a query. Based on the mutations, the two new unusual sequences could not possibly be derived from B.1.177 and known methodological differences between Pangolin and Nextclade made it clear that Pangolin was in error. In Nextclade nomenclature, the two new unusual sequences were designated as belonging to clade 20A, corresponding to Pango lineage B.1, a basal and diverse SARS-CoV-2 genotype. We will refer to this lineage as lineage X until its official designation in the story.

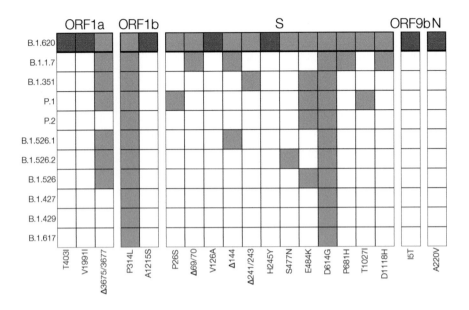

Figure 5.9 Mutations unique to the mystery lineage X (which became B.1.620) are shown in red, blue highlights amino acid changes present in other VOCs and VOIs of the time.

To better understand this new and unusual lineage X, its constellation of mutations and deletions was examined. Most prominent amongst the mutations present was E484K in the Spike protein, which had been predicted early in the epidemic to have a significant impact on neutralising antibody titres raised against lineages without this mutation. Other mutations and deletions found in lineage X were known from other VOCs: ORF1a: 3675/3677Δ (highly convergent deletion in multiple VOCs), S: P26S (found in P.1), S: 69/70del (found in B.1.1.7), S: 144/145del (found in B.1.1.7), S: 241/243del (found in B.1.351), S: H245Y, S: S477N, S: E484K (all found in B.1.351, P.1), S: P681H (found in B.1.1.7), S: T1027I (found in P.1), S: D1118H (found in B.1.1.7), and N: A220V (found in B.1.177 and likely responsible for incorrect Pango classification of lineage X as B.1.177.57) (Figure 5.9). GISAID, originally the *de facto* global repository of SARS-CoV-2 genomes, allows the user to carry out simple queries for genomes in the database. Many of the mutations that lineage X carried were already set up as possible mutation queries because of other VOCs, and it was possible to use combinations of mutations that had not been seen in any other lineage together.

Searching GISAID for genomes that carried mutations S: E484K, S: S477N, and S: 69/70del identified 40-odd additional genomes that also carried the remaining constellation of mutations and deletions characteristic of lineage X. Two Lithuanian cases of lineage X from mid-March were retrospectively identified on GISAID in this way which were missed initially because of Pango misclassification. Other genomes on GISAID were submitted by labs sequencing in Belgium, France, Germany, Switzerland, UK, and Spain. All genomes available at the time were one or more mutations away from the common ancestor genotype, none clustered strongly according to country, and lineage X was not dominant in any of the countries where it was detected (Figure 5.10). If the origin location of this lineage had an intensive sequencing program, their sequence data would likely detect certain features, such as high prevalence of lineage X within the origin region, evidence of lineage's gradual evolution and/or rise to higher frequency after originating (e.g., shown by a phylogenetic pattern where genotypes detected in the origin region sit roughly basal to diversity seen elsewhere). None of the countries where lineage X had been found at the time fit any of these features, strongly suggesting that the *actual* origin location was not represented by the sequences on GISAID.

This suspicion was strengthened by the identification of one lineage X genome on GISAID from France that came from an infected traveler coming to France from Cameroon. Authors of other lineage X sequences on GISAID were contacted personally by email to inquire about potential travel histories of patients from whom lineage X had been sequenced. Of the authors who reported back information, a surprisingly large number of lineage X genomes (around 10) had been collected from travelers coming from Cameroon to Switzerland, France, and Belgium. Some authors could not provide travel information because their organizations had deemed travel information to be sensitive and personally identifiable information. Personal information protection is an important and sensitive topic in genomic epidemiology, and there are varied opinions and regulations defining what counts as personally identifying information. These challenges can make collating the same information fields from multiple sources challenging or impossible.

As more information was being pursued, an official Pango lineage designation was requested for lineage X. The nomenclature system devised by the Pango Network to refer to distinct SARS-CoV-2 lineages has been extremely helpful for communication between scientists. With an official designation of lineage X as B.1.620, it was possible for

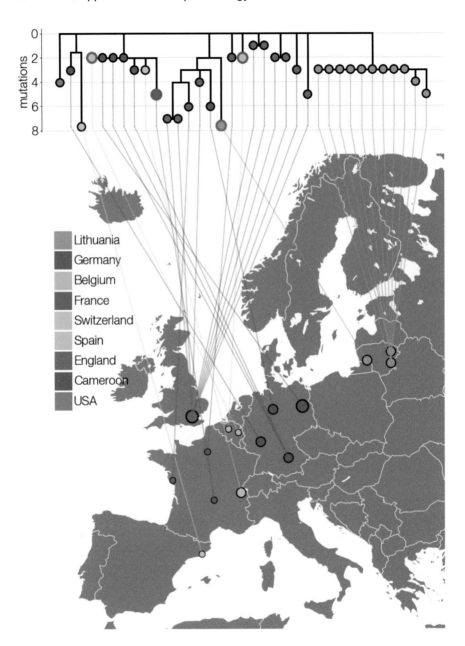

Figure 5.10 Early lineage X genomes collected from different countries in Europe. Known travel cases are indicated with outlines corresponding to country of origin.

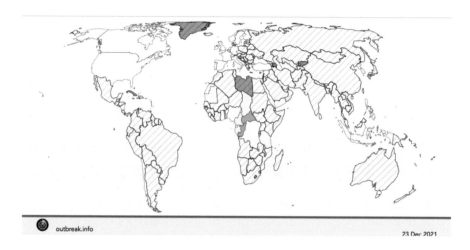

outbreak.info

23 Dec 2021

Figure 5.11 To date the highest frequency of B.1.620 genomes are observed in Central Africa (https://outbreak.info/).

personnel used to just looking at lineage composition to notice this unusual lineage, which would otherwise be assigned a different lineage. The designation also gave researchers around the world common vocabulary to discuss this lineage without having to refer to all of its mutations and deletions. An unexpected benefit of an official designation was that it tempered headlines in national and international media that referred to lineage B.1.620 (prior to designation) as an "unidentified coronavirus strain", a needlessly ominous term likely to grab attention and add to confusion when communicating with the public later.

The already strong case for an African origin of lineage B.1.620, due to the lineage's association with travelers arriving from Cameroon, became even stronger with the help of African collaborators. Scientists working in the Central African Republic (CAR) near the border with Cameroon shared six B.1.620 genomes, at last confirming that B.1.620 actually circulated in the region. As more African countries began or expanded sequencing programs, B.1.620 was eventually found in CAR, Cameroon, Gabon, Democratic Republic of the Congo, Republic of the Congo, Equatorial Guinea, and Angola. While most travelers infected with B.1.620 had arrived from Cameroon, CAR turned out to have the highest frequency of B.1.620, despite low numbers of sequenced SARS-CoV-2 genomes. Neighbouring Republic of the Congo turned out to have the second highest frequency of lineage B.1.620 in the world, though sequencing data showed up very late (Figure 5.11).

All of the evidence together suggests that lineage B.1.620 arrived to Europe from Cameroon on multiple occasions but CAR is more likely to have been the location where B.1.620 became a dominant variant, if not its birthplace. This highlights an important aspect of sequence data, namely that all available information should be brought to bear when interpreting sequence data.

Another issue that plagued the B.1.620 investigation were sequencing errors and assembly artefacts. Sequencing labs often have variable access to the tools and expertise necessary for assembling high quality genomes, and this can result in variable sequence quality. One of the labs submitting B.1.620 genomes to GISAID, for example, were able to recover genomes containing all the mutations expected of the lineage, but did not include the deletions. It would be extremely unlikely to sample a precursor lineage to B.1.620 in Europe which happened to have only the mutations but not the deletions typical of the lineage. A far more parsimonious explanation was that the reference-based assembly method used by the lab did not handle insertion and deletion changes well. Sequences like these were excluded from analyses because the presence or absence of insertions and deletions altered their phylogenetic placement.

Another intriguing case was found in Lithuanian data where a genome occupying an intermediate position along the long branch leading to all known B.1.620 genomes (Figure 5.12) was discovered weeks into the B.1.620 outbreak in the country. At face value the interpretation of such phylogenetic placement is that B.1.620 evolved in Lithuania. Given the intensity of Lithuania's genomic surveillance programme, and convincing evidence that B.1.620 had come from Central Africa, the sample was inspected in greater detail. The genome itself contained mutations typical of B.1.620 but other sites with non-reference nucleotides contained mutations typical of lineage B.1.177.60. At this point there were two hypotheses; either this was a recombinant sequence or a co-infection.

As we discuss in greater detail in Chapter 7, some of the key signals that recombination leaves were missing from this sample. For example, the mosaic pattern was only observed in this single sample. Furthermore, the actual mosaic pattern of mutations, where B.1.620-like and B.1.177.60-like mutations occurred randomly across the entire genome, did not indicate recombination. Rather, *bona fide* recombinants observed in the UK had patterns of mosaic mutations that clustered towards one or the other end of the genome.

To conclusively rule-in co-infection, the assembly files were investigated, specifically the frequencies of each non-reference nucleotide. All

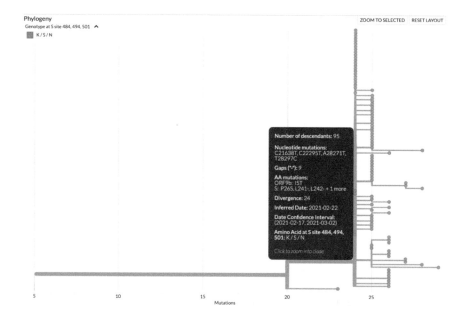

Figure 5.12   Screenshot of a Nextstrain build showing an early-branching sample (bottom) from Lithuania. This early branching sample had only some of the mutations found in B.1.620, implying that it diverged from a lineage that eventually gave rise to B.1.620, an impossibility considering all the evidence pointing to Central Africa as the region from which B.1.620 reached Europe. This sample was later identified as an incorrectly assembled genome from a co-infection case with B.1.177.60 and B.1.620 lineages.

non-reference nucleotides for that sample had a nearly 50/50 split in frequencies between B.1.620-like and B.1.177.60-like Single Nucleotide Polymorphisms (SNPs), exactly as expected from co-infection. Calling a consensus sequence from this read data, particularly when the two alternative SNPs are so close in frequency, results in the mosaic we saw initially. This sequence was subsequently removed from databases and excluded from future analyses.

An overarching lesson from B.1.620 is that, while sequencing data can provide a wealth of information, you shouldn't interpret it in isolation from all other information you may have. Sequencing errors, incorrectly labelled genomes, incorrectly annotated metadata, sequence classification tools, etc can all result in either individual or sets of observations that seemingly go against a developing hypothesis. It is therefore key to integrate as many sources of information as possible into a genomic epidemiological investigation, such that the entirety of the body of evidence is cohesive with the sequence data. And it's good to retain skepticism of sequence data that don't conform, since sequence data issues often leave imprints that are easy to diagnose.

5.4 ESTIMATING WHEN THE ZIKA VIRUS EPIDEMIC BEGAN IN COLOMBIA

Zika virus frequently causes asymptomatic or mildly symptomatic cases, which means that syndromic surveillance systems may not necessarily capture the initial few cases when Zika virus begins circulating in a new region. This issue is compounded by the fact that many other arboviral infections that circulate in South America, such as Chikungunya and Dengue virus, cause similar symptoms upon infection. Thus even if syndromic surveillance systems capture an uptick in febrile illness, this may be attributed to different viral diseases, which again may obscure Zika virus circulation. Indeed, multiple genomic epidemiological studies of Zika virus circulation in the Americas demonstrate that Zika virus was circulating for months, even up to a year, before detection by syndromic surveillance systems [15].

Why do we need to accurately date when Zika virus circulation began occurring? Firstly, this information builds our descriptive understanding of the outbreak, which is critical to informing which analytical epidemiological studies we may want to design. For example, when investigating whether Zika virus infection was related to severe sequelae such as Guillain–Barré syndrome and microcephaly, one of the sources of evidence for an association was a higher incidence of these sequelae when Zika virus infections were occurring (compared to baseline incidence rates when Zika virus was not circulating). Accurately knowing when Zika virus was circulating in a country, and therefore when individuals were at risk for Zika virus infection, is critical for selecting the correct time period for ascertaining baseline rates and ensuring the validity of such investigations.

Secondly, when mounting evidence suggested that Zika infection during pregnancy was associated with microcephaly and other congenital abnormalities, many countries began designing cohort studies to study the association in a more controlled manner, and implementing registries to provide supportive care and monitoring to families who might potentially be impacted by a Zika virus infection. In order to enroll participants in a valid way, you need to know who was at risk of Zika virus infection during pregnancy, which requires accurately knowing when Zika virus was circulating. If you underestimate the duration of time when Zika virus infections were occurring, you may errantly omit individuals from cohort studies and registries who should really be included.

So, how do we actually estimate when Zika virus circulation began? The first step is to build a temporally resolved phylogenetic tree. Here, the dataset that we use matters. Since we are interested in understanding when Zika virus arrived in Colombia, we first need to have a representative sample of the Zika virus diversity that circulated in Colombia during the Colombian epidemic. Additionally, we need contextual sequence data from other countries in the Americas that experienced Zika virus epidemics. Including sequence data from other countries has two important purposes. The first is that it allows us to see how Colombian viruses are related to viruses sampled from other countries in the Americas. This allows us to differentiate between distinct Zika virus introductions to Colombia, and to see where those seeding events potentially came from (Figure 5.13 A and B). Secondly, the inclusion of Zika sequences sampled through time from other countries ensures that we have a strong pattern of sequences available across many time points (Figure 5.13C). This serial sampling enables accurate inference of the rate of evolution, which is directly related to our estimates of when Zika arrived in different countries, which is our question of interest here.

You'll note that these trees might look slightly different than ones that you have made. Beyond having the tips coloured by country, the branches are also coloured according to country as well. This colouring is made possible by performing phylogeographic analysis in Nextstrain. Briefly, phylogeography will allow you to infer the most probable region where an ancestral virus (an internal node in the tree) likely circulated. When combined with the temporal resolution of a time tree, you can see when a geographic migration event likely occurred by looking at the estimated dates of when ancestral viruses circulated in different regions. For example, in Figure 5.14, we are showing one of the two clades of Zika virus that circulated in Colombia (and seeded transmission into other countries as well). We can estimate when Zika moved into Colombia by looking at when the internal nodes in the tree change from likely circulating in Brazil to likely circulating in Colombia.

Notably, there are sources of uncertainty here that you should consider depending on how you seek to use the timing estimates. Firstly,

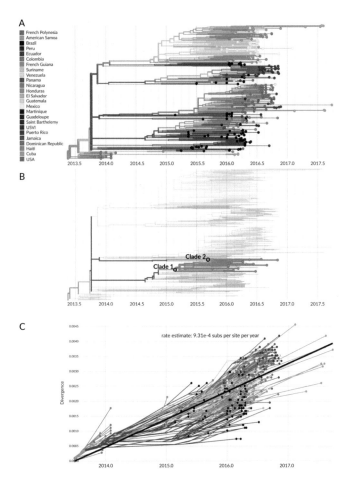

Figure 5.13 (A) A temporally resolved phylogeny of 360 whole Zika virus genome sequences, of which only 20 are sampled from Colombia. The tips and branches are coloured according to the country where they were sampled from, or the inferred country of circulation, respectively. (B) The same phylogeny as shown in the A, but with only Colombian samples highlighted. While they are pushed together in the same area of the tree, there are two distinct clades of Colombian sequences. This shows that there are two different pools of Zika viral diversity circulating in Colombia. (C) A root-to-tip plot showing the estimated evolutionary rate of Zika virus given the samples analysed here. Note the presence of sampling over time, including viral genomes sampled from French Polynesia in 2014 and from multiple countries in the Americas from 2015 through 2017.

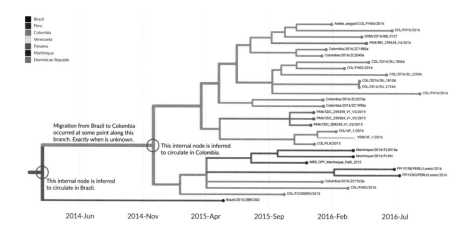

Figure 5.14 A zoomed in view of the larger Colombian clade of Zika virus within the same tree that we have been using during this case study. We have annotated two internal nodes: the most basal internal node that we infer circulated in Colombia, and then the direct parent node which we infer circulated in Brazil. This gives us a time range: on one side we have an idea of when Zika virus was likely still circulating in Brazil, and on the other end we have an idea of when Zika was likely circulating in Colombia. Therefore, the movement event from Brazil to Colombia occurred sometime between those two bounds, but we do not know exactly when. Notably, there is uncertainty on the dates for both internal nodes, which in Nextstrain are given as 95% confidence intervals around the inferred date. Capturing this uncertainty is important, since the molecular clock is an average rate that can change depending on the dataset, and the inferred dates on internal nodes can also change given the inclusion of more or less data.

there is uncertainty in the dates assigned to internal nodes. Therefore, you should always report the 95% confidence interval around the dates, and not just the point estimate. Secondly, the dates assigned to internal nodes can shift, and confidence intervals around the point estimates can widen or shrink, depending on the dataset you use. This wobble in the estimates is normal, and will come from slight changes in the evolution-ary rate when different sequences are included in the dataset, and from different numbers of sequences clustering together within the clade you are dating. Typically, the more samples you have within the clade, the more precise your date estimate will be.

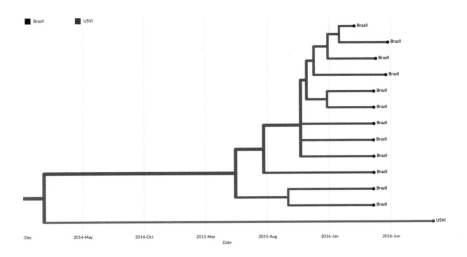

Figure 5.15 This figure shows a clade grouping multiple sequences sampled from Brazil (dark green), and one sequence sampled from the United States Virgin Islands (USVI) (pink). Note that close to two years of transmission separates the most basal ancestral node of this clade (which we infer circulated in Brazil) from when a descendant virus was sampled in the USVI. It is highly unlikely that this is a direct migration from Brazil to the USVI. Rather, this finding likely occurs due to lack of genomic sampling of the epidemic as this lineage moved from Brazil through other countries before ending up in the USVI. Since we have no genomic sampling of that intervening transmission, we cannot see it in the tree, and we end up with a very, very long branch leading to an infection sampled in the USVI.

Secondly, it's important to remember that, without more advanced phylodynamic approaches, you will not be able to infer exactly when a geographic migration event occurred along a branch. Rather, you will only have estimates of when an ancestral virus likely circulated in one region, and when a child virus likely circulated in a different region. This means that you have point estimates and measurements of uncertainty on the bounds of when a migration event occurred, but you will not know the exact timing of the movement event. In cases where there is minimal genomic sampling, and you observe long branches, it will be challenging to nail down when migration occurred (see example in Figure 5.15).

Finally, when combining dating with phylogeographic approaches, it's important to remember that the phylogeographic reconstruction will not capture any transmission through countries that lack genome sequences. Thus, in addition to showing us the challenges with getting bounds on dates, Figure 5.15 also shows us how limited genomic surveillance can hamper our understanding of where an introduction came from. In Figure 5.15, we see that the inferred ancestral sequence likely circulated in Brazil, and roughly two years later we find that a strain descended from that node was sampled in the US Virgin Islands (USVI). In this case, it is highly unlikely that we have a direct transmission event from Brazil to the USVI. The more likely scenario is that this particular lineage of Zika virus migrated through other countries that lacked genomic sampling, meaning that the lineage's journey to the USVI was not sampled, and cannot be recorded in the tree.

CHAPTER 6

Tools and Methods for Applied Genomic Epidemiological Analysis

Allison Black

Washington State Department of Health, Seattle, Washington

Gytis Dudas

Vilnius University, Vilnius, Lithuania

A T THIS POINT, you hopefully want to start conducting some genomic analysis yourself. Given that this handbook is meant to ease you into working with pathogen genomic data, we feel it's important to cover some of the bread and butter analytic approaches and pieces of software you'll likely use. That said, providing detailed discussion of software packages and how to use them is a tricky topic to cover in a textbook; methods advance, new techniques are introduced, and software is improved or deprecated faster than updated editions are published.

In this chapter, we will try to strike a balance, providing sufficient detail to help beginners learn how to approach applied genomic epidemiological analysis, while avoiding overly prescriptive recommendations that may readily change. This chapter will not provide a comprehensive review of all tools and methodological approaches available. Rather, we focus on easily-deployable and fast tools that work on the time scales needed for public health action.

 DOI: 10.1201/9781003409809-6

6.1   PHYLOGENETIC PLACEMENTS

Phylogenetic placements are named as such since they place your samples of interest onto a previously-inferred, fixed phylogenetic tree. To do this, placement methods summarise the sequences of interest according to the nucleotide changes they carry compared to a reference sequence representing the root of the tree. This tree contains information about which mutations occurred along different branches in the tree. The sequences of interest are then placed into the "optimal" position on the tree according to the mutations that the samples of interest have and how those mutations are summarised by the tree. The optimal placement may be calculated differently when using different placement methods. While there are multiple methods for assessing the quality of the placement, the tools that genomic epidemiologists currently use most frequently determine the optimal placement by means of **parsimony**.

The most parsimonious placement of a sequence onto the tree is the one that requires the least number of *additional* mutations in the sequence beyond the genetic changes summarised in the tree. Notably, there may be multiple placements of a sequence onto a tree that require the same number of additional changes, and thus have the same parsimony score. Under this framework, these placements represent equally likely placements on the tree. The number of equally likely placements thereby provides a measure of uncertainty in the placement. If there are many equally probable placements then we are uncertain which placement is the "true" placement. If the number of equally probable placements is low, then we can be relatively confident in the placement position.

At the time of writing, there were two commonly-used tools for performing phylogenetic placements: Nextclade (https://clades.nextstrain. org/) and UShER (https://genome.ucsc.edu/cgi-bin/hgPhyloPlace). While both tools perform a phylogenetic placement, the phylogenies that Nextclade and UShER place samples of interest upon are different. This means that the way the output looks will be different, and what you can learn from one program may be more challenging to learn from the other. We recommend trying out and getting comfortable with both programs, as sometimes the UShER output is most informative, while other times the cohesive picture provided by Nextclade may be more helpful.

Fasta Sequence	Size (?)	#Ns (?)	#Mixed (?)	Bases aligned (?)	Inserted bases (?)	Deleted bases (?)	#SNVs used for placement (?)	#Masked SNVs (?)	Nextstrain clade (?)	Pango lineage (?)	Neighboring sample in tree (?)	Lineage of neighbor (?)	#Imputed values for mixed bases (?)	#Maximally parsimonious placements (?)	Parsimony score (?)	Subtree number (?)
hypothetical_uploaded_sequence_1	29903	0	0	29903 (?)	0	0	11 (?)	0	20C	B.1.424	USA/MN-QDX-1663/2020 | MW190437.1 | 2020-03-17	B.1	0	2	3	1 (view in Nextstrain)

Figure 6.1 A screenshot from UShER showing the first entry in the UShER summary table for the hypothetical uploaded sequences that UShER provides. When you are on the website, you can hover over each of the column headers to learn more about them. You can see that metrics are given, and the cells are coloured according to a quality assessment.

6.1.1 UShER

UShER [16] will take your sequences of interest and place them on a megatree built from sequences sourced from various public sequence data repositories (e.g., NCBI GenBank, COG-UK data, GISAID, etc). In this way, UShER enables you to compare the genetic similarity of your samples to all other sequences available in the public repository of your selection.

UShER finds the n (where you set n) most closely related samples to your own samples of interest, and places your sample onto a small "subtree" showing the relationship between your samples and the other genetically similar sequences. This capacity is particularly useful when you want to know what sequences your samples of interest are most closely related to. When your samples of interest are highly genetically similar they will frequently end up on the same subtree. However, if your samples are not genetically similar, UShER will provide multiple subtrees for each of the samples, showing your sample of interest's placement among other publicly-available genome sequences that were genetically similar to your sample of interest.

When you run UShER, the program will return a table of metrics that summarise the placement (see Figure 6.1 for an example). These metrics will provide some information about the genome sequence that you placed (e.g., how long it was, how many insertions or deletions the genome had, how many N's the sequence had etc.). The table also provides information about the placement, including how many single nucleotide variants (SNVs) or single nucleotide polymorphisms (SNPs) the sequence had that UShER used to place the sample, which sample is the most closely related to the sample of interest, the number of maximally

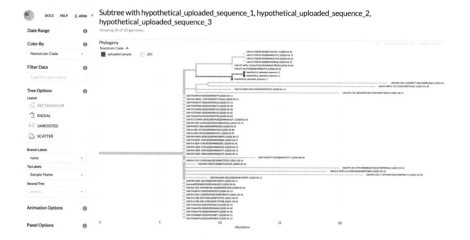

Figure 6.2 The subtree for `hypothetical uploaded sample 1` that is linked in the last column of the UShER summary table. You can see that two more of the uploaded hypothetical samples are most closely related to `hypothetical uploaded sample 1`, and so these show up on the subtree as well. You can also see that there are many different samples that have the identical genome sequence to USA/MN-QDX-1663/2020. All of the samples with identical genome sequences to USA/MN-QDX-1663/2020 are as similarly related to hypothetical uploaded sample 1 as USA/MN-QDX-1663/2020 is, so in this case you should exercise caution in your interpretation.

parsimonious placements UShER found, and the parsimony score. Of these, the number of maximally parsimonious placements and the parsimony score are some of the best metrics for understanding your placement. As discussed in the introduction to phylogenetic placements, the number of maximally parsimonious placements provides a rough measure of certainty. If there are many parsimonious placements, then you can't take any one of the subtrees as the "true placement". The parsimony score will tell you how many additional mutations were in your sequence beyond the mutations observed in other sequences and summarised by the tree. The larger the parsimony score, the more diverged your sequence is from other sequences found in public repositories. When looking at the subtree, you'll see that this parsimony score will be equal to the branch length observed between the tree and the sequence of interest. For example, in Figure 6.2 the red branch leading to `hypothetical`

uploaded sequence 1 is 3 nucleotides long, and that matches the parsimony score of 3 that you can see in Figure 6.1.

Note that while the table view in UShER will provide the name of a
sequence entry that was most closely related to your sample of interest,
there could be *multiple* sequences that have identical genome sequences
to whichever sequence shows up in the table as the "Neighbouring sample in the tree". You can see an example of this in Figure 6.2. In the
table view (Figure 6.1), you can see that the neighbouring sample in tree
for hypothetical uploaded sequence 1 is USA/MN-QDX-1663/2020 |
MW190437.1 | 2020-03-17. However, when you look at the tree in Fig-
ure 6.2, you see that USA/MN-QDX-1663/2020 is one sample among
many that have identical genome sequences, and thus stack vertically
in the tree. This means that USA/MN-QDX-1663/2020 is as close of
a neighbour to hypothetical uploaded sequence 1 as any other of
the sequences in the subtree that have identical genome sequences as
USA/MN-QDX-1663/2020.

6.1.2   Nextclade

Nextclade [17] will take in your sequences and perform some quality
assessment on the consensus genomes. You can then also place the sequences onto a background Nextstrain tree, either a globally diverse
phylogeny that is available by default within Nextclade, or some other
previously-inferred Nextstrain tree that you supply. The Nextclade phylogenetic placement has the benefit of showing a fuller picture of viral
evolutionary history leading up to where Nextclade places your sample
of interest. Additionally, the Nextclade placement will place all of your
samples of interest onto the same background tree, allowing you to see
their placements together (Figure 6.3) [16]. However, Nextclade does not
include any contextual data beyond the background tree, thus sequences
near to your sample of interest in the placement *may not* be the most
genetically similar sequences to your sample of interest. This is a key
difference in interpretation between the Nextclade placement and the
UShER placement.

6.2   PHYLOGENETIC TREES

There are two commonly used methods of tree inference: **distance-
matrix methods** and **maximum likelihood** (ML) methods. Distance-
matrix methods take pairwise distances between all aligned sequences

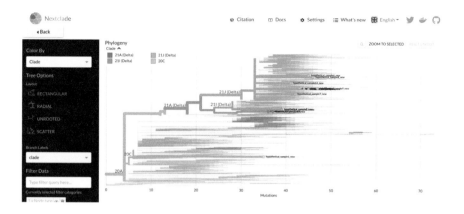

Figure 6.3 A Nextclade phylogenetic placement of 14 hypothetical samples. In contrast to the UShER subtree that we saw above, you can see the placement of all of these samples onto a single background tree, even though some of these samples are highly divergent. This provides a nice overview of the diversity of your sequences. However, this placement won't tell you which sequences are most closely related to your uploaded sequences.

and aim to reconstruct a graph that represents similarities between samples. The distance matrix methods that you'll see most commonly implemented in phylogenetic inference software are **neighbour-joining** and sometimes **UPGMA** (unweighted pair group method with arithmetic mean). While neighbour-joining performs better than UPGMA, we advise generally avoiding distance matrix methods altogether because the trees recovered by these algorithms are representations of the similarity between the samples, rather than an explicit model of how the sequences evolved. Distance matrix methods, however, are very fast for exceptionally large trees, which is why you might see them used in some cases.

In contrast, ML and Bayesian phylogenetic inference methods are more rigorous methods for inferring phylogenetic trees because they explicitly model the evolution of sequences. Thus, ML and Bayesian trees reflect the plausible relationships between sequences much better. Both ML and Bayesian methods take a long time to run on larger datasets because, unlike distance matrix methods which algorithmically cluster sequences by similarity and always give the same tree, ML and Bayesian methods search the space of all possible trees for either the single best tree (ML) or the set of most probable trees (Bayesian). That said, due to their popularity, ML methods have been iteratively improved with various approximations that make them run more quickly (RAxML and FastTree are examples).

Below, we outline some of the most common tools we see used in practice for inferring ML and Bayesian phylogenetic trees for applied genomic epidemiology. Notably, there are numerous tools for inferring phylogenetic trees, and this section will not present an exhaustive list of every tool you could use. Rather, we focus our attention on the tools that genomic epidemiologists use most frequently for public health investigations or genomic epidemiology studies.

6.2.1 Nextstrain

Nextstrain [18] is really three things:

1. An analysis pipeline for building annotated ML phylogenetic trees,

2. A codebase for interactive tree visualization within the browser, and

3. A set of pre-built interactive trees for various pathogens, maintained by the Nextstrain team.

These different components, and their different names, can sometimes be a point of confusion for new users, so we'll unpack them here. Nextstrain Augur (https://github.com/nextstrain/augur) is the analysis pipeline that takes in sequence data and performs steps such as aligning multiple sequences together, inferring a ML phylogenetic tree, inferring a molecular clock, and translating the genetic divergence ML tree into a time tree. The last step in a Nextstrain Augur pipeline is to package up all those results into a file that the user can visualise. That output file is called a JSON file, or a "JavaScript Object Notation" file. The Nextstrain JSON file will be formatted such that the file can be visualised with the second thing that Nextstrain is, a package for visualizing trees.

Nextstrain Auspice (https://github.com/nextstrain/auspice) is a JavaScript-based phylogenetic visualisation package, which you can use to visualise phylogenetic trees inferred with the Nextstrain pipeline, or any other phylogenetic tree saved as a `.newick` file. If you have inferred your phylogenetic tree using Nextstrain Augur, you can think of the JSON file as holding all of the tree information that you would like to see, and Nextstrain Auspice as the instructions for how to take that file and make it an interactive visualization that you can explore from a web browser window. To visualise a tree with Nextstrain Auspice, you can go to https://auspice.us and drop your `.newick` or your JSON file onto the browser window. This will start Auspice running on the input tree in that browser window.

Finally, nextstrain.org (https://nextstrain.org/) is a website where visitors can access interactive Nextstrain trees for various different pathogens and outbreaks, such as trees summarising the Zika virus epidemic in the Americas, the West African Ebola epidemic, or seasonal influenza circulation (among many others).

6.2.2 IQ-TREE

To create the genetic divergence tree within the Nextstrain Augur analysis, the pipeline uses an external software package that infers phylogenetic trees using a ML based procedure. If you have set up your own Nextstrain Augur pipeline, then you can choose which package you would like Nextstrain Augur to use. Currently, there are three you can choose from: IQ-TREE [19], RAxML [20], and FASTTREE [21]. By default, Nextstrain uses IQ-TREE (http://www.iqtree.org/). If you want to generate a ML genetic divergence tree outside of Nextstain, you can use IQ-TREE on its own, either using the command-line interface, or by specifying an analysis on the IQ-TREE web server (http://iqtree. cibiv.univie.ac.at/), which has a graphic user interface for specifying options.

6.2.3 RAxML

RAxML (https://cme.h-its.org/exelixis/web/software/raxml/) is another commonly used package for ML-based inference of genetic divergence phylogenetic trees. While it is highly accurate, it can take a long time to run on large datasets. While users frequently run RAxML from a command-line interface, there is a graphical user interface version of RAxML that you can download and use https://antonellilab.github.io/raxmlGUI/.

6.2.4 BEAST

BEAST [22] (https://beast.community/), which stands for Bayesian Evolutionary Analysis Sampling Trees, is a powerful tool for phylogenetic inference and genomic epidemiology. Rather than performing a ML analysis and generating a single most-likely tree, BEAST uses a Bayesian procedure that generates a posterior distribution of phylogenetic trees, which can either be summarised by a single tree, or analysed as a whole. The benefit of having a distribution of trees is that you can more effectively capture phylogenetic uncertainty than you can with a single tree. However, performing this Bayesian sampling procedure can take long periods of time (days to weeks to even months for especially large and complex analyses), especially if you lack access to a scientific computing cluster upon which to run the analysis. Therefore, BEAST is more commonly used for academic genomic epidemiological studies, and may be less useful in public health investigations that prioritise fast turnaround times.

BEAST is notable for the evolutionary models that it allows you to run, which enable you to infer how the pathogen population size is changing over time from the phylogenetic tree. This area of genomic epidemiology is often referred to as **phylodynamics** since you are analysing the dynamics of the pathogen population from the phylogeny. Phylodynamic analysis can be incredibly useful for understanding epidemiology, giving you another data stream from which to infer exponential growth rates, which can be transformed to estimates of R_0. Alternatively, you may wish to explore how enactment of different policies impacted pathogen transmission, which may be apparent from estimates of how the pathogen population size changed through time. A deeper discussion of these models and analyses in BEAST is out of scope for this handbook, but there are plenty of online tutorials and documentation for new users looking to try BEAST out https://beast.community/getting_started.

6.3 WHEN SHOULD I USE A PHYLOGENETIC TREE VERSUS A PHYLOGENETIC PLACEMENT?

Phylogenetic placements are excellent tools in many respects that provide answers incredibly quickly. Phylogenetic trees take longer to infer, but also conduct additional analyses that are not available from a phylogenetic placement. Given these trade-offs, when should you use which type of analysis?

The answer is somewhat subjective, and will be influenced by each user's particular affinity for different types of analyses and software packages, as well as the user's comfort interpreting the different types of outputs. That said, we offer the following opinions (and welcome discussion and debate!)

Phylogenetic placements are particularly useful for answering a targeted question about what other sequences your sequence of interest is most closely related to. We like phylogenetic placements for epidemiological questions in which you'd like to know, of all the data out there and available, what looks genetically related to yours. Some questions that phylogenetic placements are effective for answering include:

- What sequences are most closely related to mine?

- Could my outbreak be linked to someone else's?

- This is the first time I've detected this lineage. Where might it have come from?

- This person has a travel history. Did their infection result from exposure while traveling?

Phylogenetic trees are typically more useful when you want to explore the descriptive epidemiology of an outbreak:person, place, and time. For example, because you can translate a genetic divergence tree into a time tree, something that you cannot do currently with a phylogenetic placement, temporally resolved phylogenetic trees are preferable for looking at any sort of temporal dynamics within an outbreak, such as:

- When did my outbreak start?

- How long has my outbreak been ongoing?

- How long has this particular lineage been circulating?

The ability to also incorporate geographic data into phylogenetic trees via phylogeographic analyses also means that trees can elucidate important geographic dynamics, such as:

- How is this pathogen moving between different areas?

- Which different regions appear to be linked or show pathogen exchange?

- How frequently is this pathogen moving between these different regions?

Questions about place and time can also be explored together with phylogenetic trees. For example:

- How long did this pathogen circulate in my region before moving to a different region?

- When did this pathogen first arrive to my region, and where did it come from?

- Is my outbreak sustained by endemic transmission, or is the outbreak frequently being re-seeded from other areas?

Finally, phylogenetic trees are preferable when you have questions regarding transmission dynamics, such as:

- Did our travel policies change how frequently we saw introductions of the pathogen to our region?

- Did we manage to bring this outbreak to a halt?

- Has this lineage continued to circulate cryptically within our community, or did we manage to extinguish local transmission completely?

6.4 SELECTING CONTEXTUAL DATA FOR PHYLOGENETIC TREE ANALYSES

Selecting appropriate contextual data to include in a phylogenetic analysis can be a bit of an art, although as you do it more frequently you will begin to build up your intuition for when you lack sufficient contextual data. Often, you will need to assess your contextual data selection iteratively, building your dataset, inferring your tree, looking for areas of the tree that you believe lack sufficient context, refining your dataset, and so on. There is absolutely a need for more systematic strategies for selecting which contextual data to include, and evaluating whether sufficient contextual data have been included in an analysis. Improvements to contextual data selection are in development, but in the meantime, here are some suggestions to get you started.

Firstly, try to include external sequences that are nearly-complete and of high quality. This will usually improve the quality of multiple sequence alignment. When sequences with high numbers of N's are included in the analysis, it becomes hard to align them effectively, and those sequences can get misaligned. The effects of that misalignment can compound, with the misalignment making samples look overly divergent, which can affect your estimate of the evolutionary rate, and even lead to strange topologies within the tree.

Secondly, you will often want to include a small number of contextual samples that capture the full extent of an epidemic. For instance, even if you are seeking to describe a SARS-CoV-2 outbreak in a small town you may want to include a small number of representatively-sampled sequences from every continent where SARS-CoV-2 transmission has occurred, over the entire duration of the pandemic. These sequences help build the "back-bone" of the tree, ensuring that your analysis accurately captures the historical evolutionary dynamics that led up to your outbreak of interest. Furthermore, the inclusion of this type of contextual data can improve the precision and accuracy of your estimate of the molecular clock, since typically they will have been collected serially over a longer time period.

Finally, you will also typically want to include contextual data that are closely-related to your outbreak of interest. You will usually determine which samples fall into this bucket of contextual data by using either your general knowledge of community connectivity or by systematically selecting samples that are minimally genetically diverged from your samples of interest. As an example of the former, perhaps you know *a priori* that individuals in Community A often travel to Community B for work, and *vice versa*. In that case, it would seem reasonable that communities A and B have related outbreaks, and so you would be sure to include sequences from Community B in your analysis of Community A's outbreak. As an example of the latter, within Nextstrain analyses of SARS-CoV-2, you can select a set of samples to be your "focal" set, and then ask Nextstrain to include contextual sequences that are genetically similar to that focal set. Under this procedure, you begin with a large pool of samples from which you could draw your contextual sequences, and you select ones to include in your phylogenetic analysis based off of the number of mutations that differ between those potential contextual sequences and the sequences within your focal set.

6.5 NOTES ABOUT NODE AGES IN TEMPORALLY RESOLVED TREES

Within a time tree, the internal nodes will have estimates of their ages, and some type of confidence range around that estimate. The type of confidence interval will vary depending on whether you have used a Bayesian or ML method for inferring your time tree. Depending on what software you use, or how you've specified your analysis, these node ages may be given as:

- Calendar dates (e.g., March 07, 2017),

- Decimal dates (e.g., 2017.18),

- An amount of time beyond the age of the root (e.g., the root age is 0 years, and your sample is 1.8 years into the future from the root), or

- An amount of time into the past as calculated from the youngest tip in the tree (e.g., the most recent tip in your tree was collected on June 10, 2019, and the ancestor that you are interested in is 1.2 years back into the past from that youngest sample).

While you can readily transform between different date formats, it's worth remembering that they exist, as they can be confounding when you become used to seeing dates in one format and then encounter them in a different format.

CHAPTER 7

A Deeper Dive into Viral Genomic Complexity

Gytis Dudas

Vilnius University, Vilnius, Lithuania

T HUS FAR IN THIS BOOK, we've discussed the principles of genomic epidemiology, and how genomic diversity accrues under the assumption of clonal evolution. This is both a matter of the pathogens we've focused on up to this point – viruses – and also a matter of the techniques and analysis we've discussed – mutations accruing during the process of error-prone replication, using the rate of mutation accrual as a molecular clock, and visualizing those patterns of shared and unique mutations with phylogenetic trees. While looking at the accrual of genomic diversity through clonal evolution is fundamental to the topics and techniques of genomic epidemiology, and useful across assorted pathogens, it is not the only way in which pathogens generate genomic diversity. In this chapter, we discuss additional mechanisms by which viruses can accrue genomic diversity, and different viral genome organizations, with a particular focus on how that affects genomic epidemiological analysis and inference.

7.1 RECOMBINATION

Typically, when we talk about evolution, we consider the process by which an organism's genetic material is copied with random errors (mutations) and passed on to their progeny. Those progeny then have unequal chances of surviving and passing on that inherited variation, with additional mutations sprinkled in, to their offspring. Most probabilistic phylogenetic methods (like ML or Bayesian approaches) aim to

DOI: 10.1201/9781003409809-7

recapitulate this process by inferring whether sequences are more likely to share a mutation because the same mutation happened in other sequences, or because two or more sequences inherited the mutation from a common ancestor. The straightforward model of evolution where genetic material is inherited, mutated during replication, and then passed onto progeny is called **clonal evolution**, and it is how the vast majority of life evolves. However, recombination is another way to generate genetic variation, and it is a frequent addition on top of the process of clonal evolution.

To use an analogy, human language is often "inherited" wholesale by offspring during childhood largely unchanged, which resembles clonal evolution. Occasionally, if a child has contact with other languages, they may borrow words from that other language and intersperse it in their speech; that borrowing is similar to how recombination works. In the strictest sense, recombination happens when a strand of genetic material is broken and joined onto another. In complex organisms like animals, recombination is nearly universal and obligatory – during the production of gametes (meiosis), each chromosome we inherited whole from our parents is recombined, that is, a new chromosome is created by randomly "sampling" regions from one or the other parental chromosome. In non-eukaryotic microorganisms, recombination can integrate extraneous genetic material unrelated to anything in the host's genome in what is known as horizontal gene transfer, which is a type of non-homologous recombination (non-homologous because the host and incoming genetic material are not related). We'll talk more about this in Chapter 8. In RNA viruses specifically, nearly all recombination occurs between related sequences due to the template switching mechanism, whereby the RNA-dependent RNA polymerase dissociates from the original template it had been copying, together with the nascent RNA copy, and binds to a new template that is sufficiently complementary to the nascent copy to pair up and allow the continuation of RNA synthesis. The resulting RNA copy thus ends up being part original template and part new template.

Template switching likely takes place during all RNA virus infections, with the exception of negative sense single-stranded RNA viruses which effectively do not recombine. But in most cases, template switching will occur in a genetically homogenous infection in which the old template and the new template will be essentially the same genotype. Thus even with the process of template switching occurring, the resulting recombined sequences will not be distinguishable as such. This means that if you detect recombination in RNA viruses, likely the host had a genetically diverse infection, such as being co-infected with multiple different strains of a virus. Tools for detecting recombination are readily available but are not always well-explained or easy to interpret because they provide single numerical values like p-values, statistics, etc. Tests like these have a place in confirming recombination, but should not replace manual investigations of your sequence data for tell-tale signs of recombination.

What are these tell-tale signs? Firstly, you will likely observe an excessive number of repeat mutations; we refer to these as **homoplasies**. Walking through a toy example, imagine that at some point a virus has an A to G substitution at site 201 of its genome (A201G). If evolution proceeds only clonally, then we would expect that A201G mutation to occur and be passed on to the offspring. Notably, those offspring are descended from the same ancestor, and so they'll share other parts of the genome as well. This means that the progeny that inherit the A201G mutation will likely cluster together closely in a phylogenetic tree, and a single instance of the A201G mutation that is then inherited by the group of related viruses will explain the phylogenetic pattern. However, if recombination is occurring, then a genetically diverse set of viruses could "pick up" the genetic material carrying the A201G mutation. On a tree, this would look as though many diverged groups of viruses all have that A201G mutation, even though the rest of their genomes are quite different. If the model strictly assumes clonal evolution, then the model's only explanation for this A201G pattern across all of these different groups of viruses is to infer that the A201G mutation happened many times independently. And thus, we have excessive repeat instances of A201G: a homoplasy. For another example of homoplasies, we can return to our language analogy from earlier. Many languages, even non-Romance languages, have Latin words. If we assumed that language only evolved clonally, then we would look at these Latin words across all of the different languages, and we would say that they had been invented multiple times independently (this would be the homoplasy). Of course, the truth is that Latin words were not invented over and over again, but rather invented once in the Romance language family, and transferred by adoption to non-Romance languages.

Another feature of recombination is the spatial clustering of homoplasies along the sequence. This clustering occurs because recombination typically transfers stretches of genetic material, not just a single site. If the donor material is sufficiently diverged from its homologous counterpart in the recipient, then that stretch of genetic material will contain numerous mutations.

When unaccounted for, sufficiently recombined sequences used in conjunction with strictly clonal models of evolution lead to nonsensical inferences that can mislead the researcher or molecular epidemiologist. Judging how much recombination is too much, and when inferences are sabotaged beyond utility, is not easy and varies from situation to situation. Knowing the biology and ecology of the disease helps. If the pathogen is genetically homogenous, as is the case early in pandemics for example, or if co-infections are known to be rare, it can be argued that recombination is unlikely without even looking at the data. Other clues can also distinguish situations that produce similar signals to recombination from actual recombination. Recall that analysing recombinant sequences in phylogenetic models that are strictly clonal results in algorithms inferring that mutations transferred via recombination are recurring independently (homoplasies). While the recurrence of any given mutation (particularly in larger genomes) seems unlikely, the reality is that so many viral and bacterial progeny are produced during an infection, almost every possible mutation will occur. Thus in the presence of strong selective pressures to solve similar problems, recurrent changes are entirely expected. One of the best examples of this is E627K replacement in the PB2 protein of avian influenza A viruses jumping into humans. Virtually all human cases have this particular mutation, and with sufficient sampling of the virus in birds, it can be shown that this mutation occurs *de novo* in every single new jump into humans. Only recently was it shown that a specific interaction of the PB2 protein with ANP32A, a host protein that is sufficiently different between birds and humans, is at the root of this extremely strong selective pressure. In contrast, it is uncommon for synonymous mutations to be under similar selective pressures, and thus synonymous mutations are unlikely to recur and are a more believable marker of recombination.

One of the most thorough, unambiguous, and elegant demonstrations of recombination was performed by Jackson and colleagues [23] during the SARS-CoV-2 pandemic. Due to a comprehensive and long-running genomic surveillance programme in the UK, it had been possible to identify SARS-CoV-2 genomes assigned to lineage B.1.1.7 (Alpha in WHO nomenclature) but which didn't have all the mutations/Single Nucleotide Polymorphisms (SNPs) that B.1.1.7 should have. A closer examination of such genomes revealed that regions missing B.1.1.7 mutations had mutations typical of other lineages, strongly suggestive of recombination.

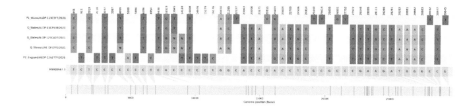

Figure 7.1 This condensed alignment shows mutations in the SARS-CoV-2 genome compared to the Wuhan-Hu-1 reference strain (shown in grey at the bottom, with mutation positions mapped below). The first and fifth sequences in the alignment (P1 Wales, and P2 England) are the non-recombinant "parent" sequences, while sequences two, three, and four (all Q Wales) represent the putative recombinant sequences. The P1 Wales virus is assigned Pango lineage B.1.177 and the P2 England virus is B.1.1.7. Looking at the alignment, you can see that the Q Wales sequences have SNPs characteristic of the B.1.177 virus in the first portion of the alignment, and then SNPs characteristic of B.1.1.7 in the latter portion of the alignment. The shift in which parent the Q Wales sequences look like, aka the recombination breakpoint, occurs between site 21255 and site 22227.

Furthermore, at least four distinct types of recombinant genomes were found, each with multiple sampled genomes sequenced by different groups. This is an extremely useful finding since it suggests that this signal of recombination did not originate from sample contamination or sequencing error. Another important aspect demonstrated by the researchers in this study was that putative parental lineages co-circulated at the same time in similar geographic locations.

One quick test for recombination is to look at multiple alignment columns that only contain two alleles within the column, but have four possible haplotypes between *all* the columns. For example, in Figure 7.1, let's take two columns of the alignment: the column at site 21255 (which contains either a C or a G), and the column at site 23063 (which contains either an A or T). These two columns span the recombination breakpoint. If we look at these two sites in the reference genome (grey), we see that the reference has 21255G and 23063A. The P1 Wales genome (B.1.177) has 21255C and 23063A. The P2 England genome (B.1.1.7) has 21255G and 23063T. The recombinant sequences (all Q Wales) have

21255C and 23063T. Thus, across the sequences we have all four possible haplotypes: GA, CA, GT, and CT. This pattern should cue us to the presence of recombination.

The logic of this simple test is that, when evolution only proceeds clonally, then one mutation arises first and the second mutation can only arise upon a genetic backdrop with the first mutation. This means that, if evolution is only occurring clonally, you shouldn't see an instance where the "second" mutation occurs in the absence of the "first".

7.2   SEGMENTED GENOMES AND REASSORTMENT

Several prominent groups of RNA viruses possess genomes that are segmented, meaning that their genomes are distributed across more than one piece of RNA. Groups with segmented genomes include orthomyxoviruses (influenza viruses with seven to eight segments being the best known), arena- and bunyaviruses (two to three segments, famous members include pathogens Lassa fever virus and Hantavirus), reoviruses (up to 11 segments, includes rota- and bluetongue viruses) and many others. Having a segmented genome can complicate many analyses in two ways.

Firstly, if you are performing metagenomic sequencing to characterise a novel segmented virus, it is usually easier to identify segments that are more conserved by homology, such as RNA-dependent RNA polymerases. It is usually much harder to reconstruct smaller and less evolutionarily constrained segments that do not resemble other sequences available on public databases. This problem may be circumvented if there are multiple independent samples of the same virus where the consistent co-occurrence of closely related stretches of sequence may be used to identify likely viral segments. Once these portions of the genome are identified as potential segments, it is not uncommon to carry out rapid amplification of cDNA ends (RACE) on putative segments to strengthen the hypothesis that they might be real segments. Many segmented viral groups have similar and partially complementary untranslated regions at the ends of each segment that help with genome packaging, and the presence and similarity of these may indicate their identity as segments belonging to the same genome.

The second complication associated with segmented genomes is reassortment. Reassortments can happen when two related but distinct segmented viruses co-infect the same cell. In the absence of incompatible genome packaging signals, new virions may package some combination of segments derived from both parental viruses. This often means that each segment evolves partially independently from its companion segments and the phylogenetic trees may differ markedly from segment to segment in a given collection of genomes. In some genomes (such as reoviruses), recombination may occur on top of reassortment within segments, complicating phylogenetic analysis even further. Fortunately, *bona fide* within-segment recombination is thought to be extremely rare amongst commonly encountered pathogens with segmented negative sense single-stranded RNA genomes.

7.2.1 Considerations for Analysis

When it comes to analysing reassortant data, the most common approach is to infer the phylogenetic tree of each segment independently and at least initially compare them visually. This is entirely sufficient for rare and impactful reassortments (e.g., identifying reassortments giving rise to pandemic influenza A viruses). Tanglegrams are a visualization method frequently employed to make this visual comparison easier. In a tanglegram, typically two segment trees will be plotted with their tips facing each other and the tips that belong to the same genome connected with lines (Figure 7.2). The extent to which these lines criss-cross each other is taken as a proxy for the extent of reassortment. Keep in mind however that the y-axis position of any given tip in a tree is meaningless and that all of the relevant information (including the presence/absence of reassortment) is provided by the hierarchical nesting of the clades. It is possible to get phylogenetically incompatible trees that exhibit no criss-crossing of tanglegram lines because of how the clades are oriented! An extension of tanglegrams is called a tangled chain where multiple segment trees are typically plotted facing one way with tips belonging to the same genome connected using lines (Figure 7.3).

Rigorous methods for reassortment inference exist and vary in terms of their computational demands and performance. GiRaF [24] is an older method that infers reassortments from posterior distributions of segment trees inferred using Bayesian methods and reports clades that are reassortant and which segments were involved. TreeKnit [25] is a recently developed method that takes in individual segment trees and reconstructs an **ancestral recombination graph** (ARG), a data structure that in addition to familiar splitting events also includes merging events that indicate reassortments and show what reassorting segments are most closely related and which corresponding segments were overwritten in the recipient genome. Another similar method that is particularly robust

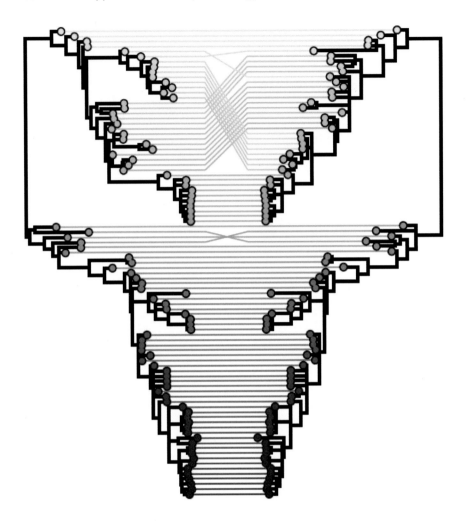

Figure 7.2 A traditional tanglegram. Tips belonging to the same genome are connected between a pair of trees inferred from different segments, in this case PB1 and PB2 segments of human influenza B viruses. For the most part the two trees resemble each other closely with the exception of a clade (yellow–green) at the top that has reassorted in the right tree with respect to the left.

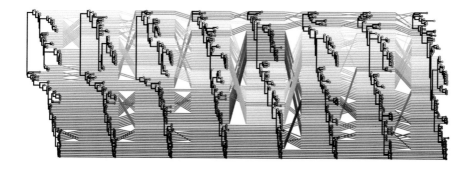

Figure 7.3 An example of a tangled chain. Here, all eight segment trees that comprise the human influenza B virus genome are plotted facing the same way and tips belonging to the same genome connected with lines.

but intensive is the CoalRe [26] package for BEAST2 [27]. It is intended for molecular clock analyses and in addition to recovering the posterior distribution of ARGs also returns the embeddings of each segment in ARGs, that is, the inferred time tree of each segment. An extension to the CoalRe method that can also reconstruct discrete character states (e.g., locations) is called SCORE.

7.3   HYPERMUTATION

Hypermutation is not intentionally a method by which viruses generate genetic diversity; it is traditionally a host-defense mechanism, whereby the hypermutation should hopefully change the sequence so greatly as to render the progeny virion non-viable. However, hypermutation doesn't always kill off a virus, and we can observe tracts of hypermutation in genomic datasets for certain pathogens. We will therefore discuss how to identify hypermutated regions of viral genomes, and how to handle those portions of the sequences when performing genomic epidemiological analysis.

There are a few eukaryotic enzymes that appear to be used by the host to attack viral genomes within cells. The most famous examples include APOBEC (apolipoprotein B mRNA editing enzyme, catalytic polypeptide) and ADAR (adenosine deaminase acting on RNA), both of which sometimes leave distinct patterns of changes in viral genomes that survive encounters with these proteins.

APOBECs are the more diverse of the two and carry out cytidine deamination on RNA and DNA – they turn cytosines into uracils (RNA analog of thymine) such that a C:G pair becomes a U:G pair. The U:G pair is a mismatch, which will generally be repaired to a matched T:A pair, resulting in a C to T mutation.

There are many different APOBECs that are classified with numbers and letters (e.g., APOBEC1, APOBEC3A, etc.) whose functions are varied and potentially not understood fully at this time, and range from an antiretroviral role to aiding affinity maturation of B-cell receptors.

The function of ADARs, though somewhat clearer, is also not fully understood. They target double-stranded RNA (dsRNA) which is usually a very strong indicator of an RNA virus infection (cellular RNAs typically exist as either a DNA-RNA hybrid during transcription or single-stranded mRNA, tRNA, or rRNA otherwise). The replication of RNA virus genomes produces an intermediate step that is double-stranded RNA. An ADAR enzyme recognizing this dsRNA may deaminate adenosines into inosines which behave like guanines (i.e., pairing with cytosine), resulting in A to G mutations.

When looking for putative host-driven hypermutation patterns in viral genomes, you should be aware of the nucleotide context that the enzyme in question prefers. For example, APOBEC3 has a preference for changing TC to TT, that is, the edits of C-to-T are biased if there's a preceding T. As an additional point of consideration, edits may appear in two different ways depending on whether edits are occurring on the sense or antisense strand. For example, if we were looking for ADAR edits, we should look for both A to G and T to C mutations since dsRNA can be approached (and edited) from either side.

As a concrete example, during the West African Ebola virus epidemic there were a number of Ebola virus genomes that carried stretches of tens to a bit more than a hundred nucleotides with clustered T to C mutations but not A to G (Figure 7.4). The current thinking goes that ADAR preferentially edits As into Gs on the sense strand (as opposed to the antisense, that is, the reverse complement of the genome) of negative sense single-stranded RNA viruses. Since bioinformatically we overwhelmingly work with RNA virus genomes pointing the "conventional" direction (5' to 3' with genes running left-to-right) we observe these as stretches of T->C mutations.

If you observe putative tracts of hypermutation and you plan to run molecular clock analyses on these sequences, you should seriously consider masking the sites of putative editing. Any molecular clock models

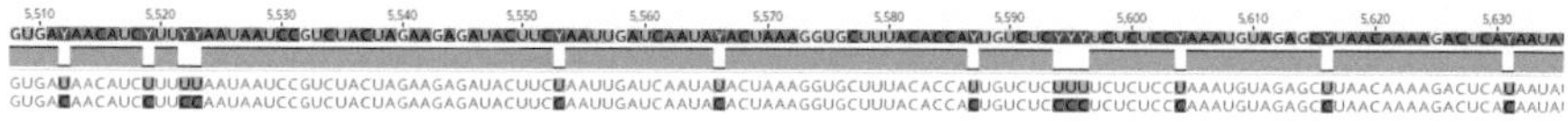

Figure 7.4 An example of a strong ADAR editing signal in the Ebola virus genome from the West African epidemic. The upper sequence (Makona-Gueckedou-C05) is an early sequence from the epidemic while the bottom one (J0169) is from an outbreak in Freetown later in the epidemic and bears a high number of T/U->C mutations in a very short stretch. There are about as many mutations in this small stretch as there are other non-clustered mutations differentiating these genomes.

the slow trickle of mutations that are introduced by the viral polymerase and that manage to survive by not being sufficiently deleterious. A sudden influx of lots of mutations (in Figure 7.4 it's roughly doubling the number of mutations there were before) the molecular clock can only assume that the overall molecular clock rate is much higher and therefore the entire tree is much younger. This is less of a problem for more sophisticated clock models, like relaxed uncorrelated clocks, but should still be addressed head-on if there are strong suspicions that questionable mutations came about via anything other than the viral polymerase making errors.

Note that we will not always have the luxury of neatly defined stretches of hypermutation. During the mpox (previously called monkeypox) outbreak of 2022, there were strong indications that the clade causing the outbreak had at some point experienced APOBEC3 editing albeit without a spatial clustering as extreme as the Ebola virus example in Figure 7.4.

7.3.1 Handling Variable Mutation Rates with Relaxed Molecular Clocks

In Chapter 2, Section 2.3, we discussed molecular clocks, and how you could estimate the evolutionary rate by fitting a linear regression line through a root-to-tip plot. This type of clock is referred to as a **strict clock**. It is considered a strict clock because it fits the same evolutionary rate to all branches in the tree.

The idea that all branches in the tree evolve at the same rate is a simplifying assumption that lowers the parameterization of the analysis and generally makes computational analyses run faster [22,27]. Strict clocks often work well, but there may be times when you want to capture evolutionary rate heterogeneity across the tree. To do so, you need to estimate branch-specific evolutionary rates, and we can do this using **relaxed clocks**. One common method for estimating a relaxed molecular clock is using BEAST or BEAST2. While there are multiple ways to specify the relaxed clock, most commonly we consider the rate heterogeneity to be log-normally distributed and uncorrelated.

What does this mean? Firstly, branch-specific evolutionary rates are drawn from a log-normal distribution, which is a probability distribution that looks like a normal distribution in log-space. The draws are considered uncorrelated because each branch gets an independent draw regardless of where the branches are located in the tree, as opposed to forcing parent-offspring branches to have some correlation in their rates. During analysis, the parameters of the log-normal distribution, the mean and standard deviation, that best fit the sequence data at hand are estimated. While you can look at the branch-specific rates, the mean rate across the entire tree is computed by summing all the mutations that must have happened across the tree (computed as the evolutionary rate for each branch multiplied by the length of that branch) and dividing that sum by the sum of all branch lengths in the tree.

Why might you want to use a relaxed clock? Firstly, using a relaxed clock prevents branches with more or fewer mutations than the average from having an undue influence over the molecular clock rate across the entire tree. Essentially, it prevents the average evolutionary rate from being unduly influenced by outliers (such as might occur if you haven't masked tracts of hypermutation). Secondly, sometimes branch-specific rate variation is one of the specific quantities that you would like to measure. For example, when investigating cases of sexual transmission of Ebola virus disease from survivors, a common indicator that such transmission has occurred is that the recipient viral infection looks genetically quite similar to the strain that the infector was infected with, even when many months have passed. On a tree estimated under a relaxed clock, this relative lack of genetic divergence given the amount of time that has elapsed between the infector's infection and the infectee's infection will show up as a much slower branch-specific evolutionary rate. For examples, see Mbala–Kingebeni and colleagues' investigation of an Ebola virus disease relapse infection [14], or Rambaut, McCrone, and Baele's (2019) tutorial on using relaxed clocks in BEAST to estimate Ebola virus rate heterogeneity (https://beast.community/ebov_local_clocks.html).

7.4   DIVERSITY OF RNA VIRUS GENOME ORGANIZATIONS

There will be many cases where you will follow standards of analysis set by the pathogen's research field. However, if you're ever in a position where you encounter a new pathogen or decide to venture outside your comfort zone and explore other pathogens, you might encounter some evolutionary solutions viruses have come up with that may affect the way you carry out your analyses. You should be aware of these unique edge cases and consider how your methods might be affected by such sequences. The following is by no means an exhaustive list of considerations but one that might give you an appreciation for what to be on the lookout for.

7.4.1   Ambisense

Though for the most part many RNA virus genomes code their genes in the same direction (i.e., RNA polarity – 5' to 3' or 3' to 5'), there are some groups that vary in this regard. One extreme example (and not relevant for human public health at the time of writing) are narnaviruses, some of whose members contain two open reading frames running in opposite directions that overlap across the entirety of their segment(s). Only one of these is thought to be functional. Human-relevant viruses with ambisense genomes include arenaviruses (e.g., Lassa and Machupo viruses) where each segment contains a reverse-complementary loop sitting between a pair of genes each "pointing" towards the loop. If you are working at the nucleotide level, further analyses might not be affected, but you should keep these nuances in mind if you decide to work at the amino acid level and are not using programs that support annotations and their automated extraction/translation from nucleotide sequences.

7.4.2   Reverse Complementarity

At some point you may encounter viruses that use RNA structures to complete their life cycles. Hepatitis delta virus, the causative agent of Hepatitis D, is a satellite virus that can accompany hepatitis B virus infection. Hepatitis delta virus has a circular single-stranded RNA genome which is highly self-complementary and so adopts a partially double-stranded RNA structure. Another example of reverse-complementarity in RNA viruses includes segment ends that are used for genome packaging.

The easiest way to check whether such sub-sequences exist in a given genome/segment is to run self-similarity dot plots. There are a number of software packages that can do that for you (e.g., dotmatcher). They will plot a pair of sequences (can be same or different), one along the x-axis and one along the y-axis, and put a dot for every nucleotide that matches. If plotting a sequence against itself, you should see a diagonal, one-to-one, line running through the plot (since the sequence is identical to itself at each position). Any off-diagonal elements indicate repeats, and if the method allows for complementarity, any offset diagonal lines oriented the opposite to the diagonal one-to-one line will indicate a region that is reverse complementary to another region.

7.4.3 Splicing

Sometimes viruses may process their mRNAs via splicing, which can present a challenge for interpreting sequencing data and/or evolutionary analyses. Influenza A viruses are famous examples of this where transcripts of at least two segments (matrix and non-structural) are spliced before being translated into at least four proteins with distinct functions. Splicing may be suspected in RNA virus genomes if there appear to be regions with excess non-coding space. For many RNA viruses, such genomic space is a premium resource and so must be explained. You can strengthen your suspicions by looking at whether long open reading frames offset from the "main gene" exist nearby, and get evidence of splicing by sequencing infections to a great depth since occasionally you will be able to catch both the genomic unspliced and spliced transcript RNAs. Note that splicing happens in the eukaryotic nucleus and thus might not apply to the virus you're working with.

CHAPTER 8

Genomic Epidemiology of Bacteria

Taj Azarian

University of Central Florida, Orlando, Florida

Allison Black

Washington State Department of Health, Seattle, Washington

IN OUR DISCUSSION of genomic epidemiology up to this point, we've focused on the accumulation of mutations across the *genome*. While this is common terminology, at this point in the discussion it's worth clarifying that we have specifically been discussing the accumulation of mutations along the *chromosome*. For viruses, the chromosome is the only genetic element that they contain. That chromosome can be one or multiple strands of nucleic acid, either RNA or DNA, and may be either single- or double-stranded. The chromosome encodes the genes for making proteins essential for the virus to replicate and assemble functional progeny viruses.

Like viruses, bacteria have chromosomes, but they can carry additional genetic elements as well. Certain genes or entire genetic elements may be present in some but not all bacteria within a species. Still some genetic elements can be present in different bacteria *across* different species. These patterns of gene gain, loss, and exchange make the bacterial genome complex and dynamic. In this chapter, we'll break down the different types of genetic loci that contribute to the bacterial genome, and how we tend to use those different genomic elements in genomic epidemiologic analysis. Readers focusing on designing, analysing, or interpreting genomic epidemiological analyses of bacterial diseases should benefit from this chapter.

8.1 BACTERIAL MECHANISMS FOR GENERATING GENETIC DIVERSITY

Like viruses, bacteria accrue mutations during genome replication. However, unlike RNA viruses whose RNA-dependent RNA polymerases *lack* proofreading capability, the DNA polymerases that bacteria use to replicate their genomes *do* have proofreading ability. This means that when the wrong nucleotide is incorporated during genome replication, that error can be fixed. This leads to one of the key differences between viruses and bacteria for genomic epidemiology; bacteria have much slower evolutionary rates compared to viruses. For example, bacterial genomes tend to accrue substitutions at a rate of 1×10^{-8} to 1×10^{-5} substitutions per site per year [28], while RNA viruses tend to exhibit evolutionary rates between 1×10^{-4} to 1×10^{-3} substitutions per site per year [10]. Even when accounting for differences in genome length, a hypothetical 15,000 nucleotide long viral genome with an evolutionary rate of 1×10^{-3} substitutions per site per year may accumulate 15 substitutions across the genome in a year, while a five million base pair long bacterial genome with an evolutionary rate of 1×10^{-6} substitutions per site per year will accumulate five substitutions across the genome in a year.

Since our power to investigate infectious disease epidemiology using genomic data comes from the overlapping timescales of genome evolution and infection transmission, having mutations occur less frequently reduces our ability to resolve transmission, especially on short timescales. Indeed, when a bacterial strain causes an acute outbreak, many if not most of the outbreak-associated cases will have identical or nearly identical core genome sequences. In such cases, genomic analysis sensitively defines which cases form part of the outbreak, but may not provide sufficient information to infer *within-outbreak* transmission dynamics. For these applications, the primary advantage of whole genome sequence data will be to rule out sporadic cases from an outbreak cluster.

Beyond just their size and their slower evolutionary rates, the practice of bacterial genomic epidemiology is complicated by the fact that sequence diversity is not simply generated through the accumulation of single nucleotide polymorphisms that are inherited by offspring, but also through the loss and acquisition of genetic elements through horizontal gene transfer (HGT) between other bacterial "peers". We introduce these mechanisms in more detail below; however, one important note is that sequence polymorphisms derived from mechanisms *other* than clonal evolution should not be used in phylogenetic analyses, as these additional sources of genetic diversity can confound or obscure both the molecular clock and the relationships between isolates. This ability puts the analyst in the tricky situation of needing to determine which SNPs present across the genome occurred through clonal evolution and which occurred though other mechanisms and must be masked from the analysis. While software exists to conduct such decision-making programmatically, the need to know whether the pathogen of interest typically evolves clonally (such as *Mycobacterium tuberculosis*), or whether diversity might be generated through recombination (such as with many bacterial species including *Streptococcus pneumoniae*), requires familiarity with your bug, as well as additional effort upstream of your genomic epidemiologic analysis to prepare the dataset.

While we have discussed clonal evolution and the vertical inheritance of mutations in this book already, we will now introduce how genetic diversity can be generated via the uptake of DNA from external sources, whether that be from other somewhat closely related bacteria, different species of bacteria, or the environment, via HGT. Bacteria may inherit small tracts of DNA of a few hundred nucleotides of length, or may inherit large elements encoding multiple genes. Some horizontally acquired genetic sequences integrate into the bacterial chromosome, while others can replicate stably and independently from the chromosome once inside the bacterial cell. Acquisition of this DNA can result in minor or significant genome sequence change if that DNA is integrated into the chromosome, either through **homologous** or **non-homologous recombination**. Alternatively, the acquisition of horizontally transferred elements may only change the accessory genome content of the bacterial isolate.

In this section, we introduce these different processes, how they impact genetic variation observed through whole genome sequencing, and how these processes should be used, considered, or corrected for when leveraging bacterial genomic data to support public health investigations.

8.1.1 How Does Horizontally Acquired DNA Get into the Bacterial Cell?

Before horizontally acquired DNA can change the genome of a particular bacterium, it must first be brought into the bacterial cell. There are a myriad of ways in which DNA uptake can occur, with plenty of exceptions and special cases. For the purposes of this text, and this audience, we will not exhaustively review the molecular microbiology of these mechanisms. If the reader is interested though we encourage further study.

There are three primary ways in which a bacterium can uptake external DNA. Firstly, bacteria can take in DNA present in the environment through a process called **transformation**, in which DNA is brought in through a pore in the bacterial cell membrane. Secondly, in a process called **transduction**, the genome of a bacteriophage (a virus that infects bacteria, also shortened to phage) is brought into the bacterium upon infection. Some species of phage have genomes that integrate into the bacterial chromosome (lysogenic phage), while other phage species (lytic phage) will simply use the bacterial cell for viral genome replication, package the viral genomes into progeny phage, and kill the bacteria releasing the progeny. The final common mechanism of DNA transfer is **conjugation**. During conjugation, an apparatus called a pilus acts as a bridge connecting two bacteria directly, facilitating DNA transfer from the donor to the recipient bacterium.

Once inside the cell, what happens to this externally acquired DNA? It may either integrate into the bacterial chromosome, or remain outside of the chromosome. When integration occurs, this single event can lead to either minor or major changes in the nucleotide sequence, depending on the processes governing that integration event. We'll summarise those processes at a high level in the following sections.

8.1.2 Integration of Horizontally Acquired DNA into the Chromosome

Horizontally acquired DNA often integrates into the chromosome via **homologous recombination**, a process by which a typically short portion of a bacterium's genome is replaced by a sequence from a different bacterial lineage when the incoming DNA is similar (homologous) to the chromosomal DNA.

Homologous recombination is an important and fundamental mechanism for bacteria to generate genomic diversity; it is every bit as much a record of evolutionary history as the accumulation of substitutions that arise due to error-prone replication. However, it is a much harder evolutionary signal for us to interpret from an epidemiological perspective, and genomic epidemiologists typically discuss it as an obstacle that must be overcome rather than a signal to investigate. We'll clarify why this is, and how we commonly approach dealing with this natural process.

As we've discussed, clonal evolution is the process by which mutations occur during error-prone genomic replication, are integrated into the sequence, and inherited by descendent organisms. Clonal evolution typically occurs at a steady rate (the molecular clock), and introduces these mutations relatively uniformly across the genome. In contrast, during the process of homologous recombination, a small fragment of similar, but not identical, sequence replaces the homologous area of the genome that was previously present. This event has the effect of introducing *multiple* changes to the genome sequence during a *single event*, essentially accelerating the incorporation of polymorphic sites in the genome compared to the frequency that those substitutions would have arisen through clonal evolution alone.

The incorporation of these multiple substitutions in a single event isn't the sole problem; indeed, if these recombination events occurred a regular intervals and introduced similar numbers of genetic changes per recombination event, we'd recapture a steady rate at which genome diversity accumulates, albeit through two processes (recombination and clonal evolution). We are not that lucky. Homologous recombination appears to occur with irregular frequency, and the number of genetic changes, or even additions or deletions of whole genes, that occur during a single recombination event is highly variable. Furthermore, the length of DNA incorporated into the genome (referred to as the "track length") can vary tremendously, from hundreds to tens of thousands of base pairs. These factors thereby impact the correlation between the passage of time and the acquisition of substitutions across the genome, undermining our ability to accurately estimate evolutionary rates.

If the challenges imparted to estimating the molecular clock weren't tricky enough, homologous recombination, by the very fact that it introduces a different piece of genome into the recipient bacteria, creates a situation where different parts of the genome have different evolutionary histories. Sequences introduced through homologous recombination can make a genome look more diverged from other sequenced samples in an analysis *or* make sequences look less diverged, and more closely related, than they really are. In essence, different portions of the genome will be more or less related to other sequences – the evolutionary history written into the genome is different depending on which portion of the genome you're looking at. This reality undermines our ability to infer clear and consistent relationships between analysed samples, and the phylogenetic trees which are usually the object that we use to describe those relationships.

These challenges explain why we often treat recombination as a nuisance to weed out rather than an evolutionary process to study. So how do we handle the presence of recombination? Typically, we seek to remove sequence polymorphisms that we believe are attributed to recombination, leaving us to analyse only those substitutions in the genome that we believe occurred due to clonal evolution. Typically, we use statistical methods to detect recombinant portions of the genome characterised by a high density of substitutions within a limited sequence fragment length. Those areas will be masked, leaving us the vertically inherited SNPs to use in our phylogenetic analyses.

Importantly, this method of detecting recombination really only works with closely-related sequences, where the receipt and integration of a new genomic fragment results in an apparent abundance of mutations that is easily distinguished from the frequency of mutations in surrounding areas of the genome. When analyzing highly diverged sequences (e.g., a global dataset of a particular bacterial species), sequences will hold evidence of recombination receipt *and* donation, and recombination can make sequences look more divergent *or* more similar to each other than would be apparent from clonal evolution alone. This makes inferring the deeper historical relationships between bacteria exceptionally difficult. However, from the applied genomic epidemiology perspective, we're fortunate to generally investigate infection transmission on the timescale of days and months. The genomic diversity of an outbreak-attributable strain is typically very low within the outbreak, and recombination detection is possible.

8.1.3 Acquisition and Exchange of Extrachromosomal DNA

Thus far we've discussed how integrating externally acquired DNA into the bacterial chromosome can introduce new polymorphisms to the genome. However what happens when that externally acquired DNA *does not* integrate into the chromosome? Here, we'll discuss the impact of plasmid transmission on gene content and function, with a particular focus on public health impacts and genomic analysis.

8.1.3.1 *Plasmids*

Plasmids are sufficiently important from a public health perspective that we will discuss them in detail. Plasmids are extrachromosomal DNA elements, usually, but not always, made up of circular double-stranded DNA, and they harbour genes encoding proteins that might provide benefit to the bacterium in certain environments. They do not encode the most essential genes that the bacterium needs for replication. Rather, you can think of plasmids as carrying "bonus material" for the bacterium. Plasmids can range in size from 2 to 5 kilobase pairs (Kbps) to up to 500 Kbps.

Most commonly, plasmids replicate independently of the chromosome, although some plasmids can integrate and excise themselves from the chromosome, and when integrated get replicated during the process of chromosomal replication. Bacteria can gain and lose plasmids, and they can harbour more than one plasmid, although certain plasmids may not be able to stably co-exist within a bacterium, a feature known as **plasmid incompatibility**. Furthermore, plasmids can be transmitted between different bacteria, both within and between bacterial species, via HGT.

8.1.3.2 Analysis

Depending on the sequencing and genome assembly approach (discussed in more detail later in this chapter), plasmids found in bacterial strains can be partially or fully assembled. Oftentimes, review of the annotation files will show phage-associated coding sequences, and depending on the species, plasmid "typing" tools exist to identify well-known plasmids. This identification is usually made based on the gene content of the plasmids. When carrying out analysis on a large sample of bacterial genomes, one common component is to identify the presence and absence of plasmids or at least the epidemiologically important genes such as antimicrobial resistance determinants, that are found on them. This can be particularly important when investigating plasmid-mediated outbreaks. For these investigations, tracking the plasmid may be the primary focus of the investigation, even in instances where the bacteria harbouring them are different.

8.1.3.3 Limitations

Plasmid assembly can be hard using short read sequence data because portions of the plasmid may have sequence homology with the chromosome or even contain repeat regions within the plasmid. In addition, plasmids are often present in high copy numbers in relation to the bacterial chromosome. Together, this complicates the assembly process, making it hard to finish plasmid sequences. Long read sequencing and hybrid assembly, discussed in Section 8.5.1, can sometimes be used to obtain complete plasmid assemblies.

8.2 A BRIEF NOTE ABOUT BACTERIAL POPULATION STRUCTURE

Several chapters could be spent discussing how and when bacteria get separated into subpopulations that evolve distinctly from each other (population structure) and bacterial speciation. Simply, the population structure of bacterial species is complex. As described, genetic variation in bacteria is driven by recombination and mutation, and that variation is acted upon by selection and chance. Combined with other biological and ecological factors, these processes can shape the structure of a species or blur the boundaries between them. This process is ongoing and large changes usually occur over longer evolutionary timescales.

Thankfully, outbreak investigations mostly focus on a subpopulation of a bacterial species over shorter timescales. How structured a bacterial population is ranges from completely unstructured (termed **panmictic**) to well-structured (or **clonal**). Where a species falls on this spectrum is measured by how much statistical association is found among loci on the genome, referred to as **linkage disequilibrium**. Without diving into a detailed discussion on bacterial population genetics, the important concept is that the more recombination a species experiences, the more mixed it becomes, and the less structured the population [29]. This of course is an oversimplification. In practice, few bacterial species are regarded as truly panmictic; *Helicobacter pylori* and some environmental species of *Vibrio* are included in that group. *Staphylococcus aureus* and *Mycobacterium tuberculosis* are canonically referred to as clonal, while most others, such as species of *Neisseria* and *Streptococcus*, fall in the middle. When investigating a putative outbreak, it is important to have a general knowledge of the suspected species' population structure.

8.3 BACTERIAL SEQUENCE TYPING

Prior to the advent of whole genome sequencing, many serological and molecular based typing methods were developed to define bacterial population structure and classify strains, several of which are species specific. These include multi-locus sequence typing (MLST), pulsed-field gel electrophoresis (PFGE), *spa* typing for *S. aureus*, restriction fragment length polymorphism (RFLP) analysis, serogrouping/serotyping, and others. The most common, MLST, uses allele profiling in seven housekeeping genes to define the sequence type of a particular bacterial strain. Each bacterial species has a specific MLST scheme, which can be found at https://pubmlst.org/. For each MLST scheme, the particular nucleotide sequence of each gene is summarised with a label, which is typically a number. When another strain has the *exact* same nucleotide sequence at that same loci, they are annotated with the same number. If a particular strain has a sequence never before observed at a particular loci, then that sequence will be provided a new numerical label. This allows a bacterial strain to be summarised according to their sequence type (ST) – whereby the numerical labels that summarise each loci map back to a particular sequence motif, and make comparing sequence identity across a large genome more wieldy.

The trade-off for this easy annotation is that when your strains have *different* sequence types, the MLST scheme will not summarise the *amount* of genetic distance between your different sequence types. The MLST annotation provides a singular label to define a sequence motif across particular genetic loci. When you observe two different labels for the same loci, that information does not tell you whether those two sequence motifs differ by one or tens or hundreds of substitutions. To complicate things, a single mutation in a housekeeping gene may result in the assignment of a new ST. In addition, the same ST can wind up on different genomic backgrounds through recombination, and the same STs may even *de novo* evolve in different lineages. As with many things, the ease that comes from MLST also abstracts away some of the data richness, which you may find useful to return to when diving into bacterial genome analysis. With greater access to cost-efficient sequencing, we can now sequence larger portions of the genome, leading to the ability to define a multi-locus sequence type using all genes present in the core genome. This type of MLST, often referred to as cgMLST – core genome multi-locus sequence typing – recently rose in popularity especially for use in large genomic databases. However, for conducting genomic epidemiology studies, all variation found in the "core genome" (detailed below) is routinely used for analysis.

8.4 DEFINING BACTERIAL GENOMIC ELEMENTS

Typically, we discuss bacterial genome organization in terms of the **core genome**, the **accessory genome**, or the **pangenome**.

8.4.1 Core Genome

The core genome is defined as the set of genes that are present across all strains within a particular sample from a bacterial species. This is a functional definition as the core genome changes based on the samples you are looking at, which could include members of a single bacterial lineage, species, or even several closely related species. As genome content varies considerably even among members of the same species, the core genome, by definition, is that portion of the genome that is present and comparable across *all* samples of interest.

8.4.1.1 Analysis

Genomic epidemiological analysis of the core genome typically focuses on defining relationships between isolates according to the patterns of shared and unique substitutions, acquired through error-prone replication, observed across different isolates' core genomes. With core genome analysis we regain much of the ease of viral genomic epidemiology. Namely, by using the core genome we ensure that we're looking at genomic regions where we have information for all samples; we do not have to consider how to incorporate gene presence/absence information. And by focusing on single nucleotide polymorphisms (SNPs) accrued across the core genome through clonal evolution, we can approach genomic epidemiological analysis using the same principles as we've discussed for viruses. We can align sequences to each other, compare which sites in that alignment show sequence polymorphism, and use that information to perform phylogenetic analysis.

Since bacteria have much larger genomes than viruses, they can be harder to work with computationally. This situation means that while for viral genomic epidemiology we tend to store, share, and analyse the entire genome sequence, for bacteria we will often consolidate and summarise sequence identity, either by only looking at portions of a multiple sequence alignment where sequence polymorphisms exist, or by summarizing sequence identity with a categorical annotation (e.g., MLST, as defined previously).

The size of the core genome will decrease as the number of samples you are considering increases. The reason for this is that as the diversity – both in terms of genome content and nucleotide variation – of the sample set increases, fewer genes are shared across *all* samples in the set. An extreme example of this phenomenon would be if you were to analyse the core genome of samples from different bacterial species. The set of genes present across multiple bacterial species will be much smaller than the shared set of genes present within a particular clade of a single species. From a phylogenetic interpretation perspective, this is why dominant clades in a species-level phylogeny will often have very short terminal branch lengths. With a core genome that has been whittled down to genes present across multiple species, there's much less sequence (and therefore genetic variation) observable within the species or clade.

8.4.2 Accessory Genome

The accessory genome is all of the genetic elements, either on the chromosome or harboured on mobile genetic elements such as plasmids, phages, or integrative and conjugative elements, that are present within a particular bacterial strain, but are not found across all samples in your sample set. These genes are often epidemiologically relevant, conferring antibiotic resistance or virulence (such as biofilm formation or toxin production), and can come and go depending on availability and the needs of the organism in a particular environment. In some instances, "core genes" impacted by high rates of mutation and/or recombination can be so diverged that they cluster separately. In these cases, such as for *S. pneumoniae* genes *pspA* and *pspC*, a "variant" of that gene can be found in every member of the species, but the amount of diversity precludes the ability to align all of them. Therefore, they wind up in the accessory genome. Again, this emphasises that this system of categorization is largely functional rather than biological.

8.4.2.1 Analysis

For bacteria with dynamic genomes, meaning that they experience frequent gene acquisition and gene loss, looking at gene content variation across the accessory genome can provide an additional source of information when differentiating between cases. For example, during a hospital-associated outbreak of *Acinetobacter baumannii*, Mateo-Estrada and colleagues [30] found that many cases had identical core genome sequences, even though cases and sample collection had occurred over multiple years. However, when applying a simple assessment of whether a gene was present or absent from a sample's accessory genome, they were able to capture genomic content divergence between the different outbreak cases.

8.4.2.2 Limitations

While genome content variation provides additional resolution to observe genetic similarity or dissimilarity between cases, it may be challenging to use that data beyond a determination that cases are truly clonal (identical) or not. Specifically, it's much harder to use gene content variation to investigate transmission over time, which is how we've typically discussed evolutionary analysis throughout this book. When changes to the genome occur by chance and are inherited by descendents through clonal evolution, the degree of sequence identity observed parallels the amount of transmission that separates cases. However, when genes can be easily lost or acquired, and this process may be driven by selective pressures or the environment, it becomes much harder to assess what it means epidemiologically to share a genetic element. Specifically, do two cases share a particular accessory gene because they have epidemiologically linked infections, and one case inherited that genome organization from the primary infection? Or are the cases epidemiologically unrelated, but both occurred in a similar ecological setting – such as the use of a particular antibiotic – where the acquisition of an antimicrobial resistance gene would confer an advantage? Finally, as with all computational analysis, errors in genome assembly and annotation can result in spurious differences in gene content.

8.4.3 Pangenome

The pangenome represents all genetic elements present in a bacterial isolate – both the core genome and the accessory genome together. When genomic epidemiologic studies consider both phylogenetic patterns implied from core genome SNP data and patterns of variation in accessory genome content as information about organism relatedness, we term such analyses "pangenomic epidemiology". A notable characteristic of pangenome variation is that, due to patterns of recombination, core genome SNP diversity and accessory genome diversity (measured by the presence and absence of accessory genes) are correlated. The more distantly related members of a species are in their core genome SNPs, the more dissimilar their accessory genomes. This is somewhat intuitive if we consider population structure, as briefly described above.

8.5   BACTERIAL GENOME SEQUENCING AND ASSEMBLY

To generate whole genome sequences for viral genomic epidemiological analysis, the lab will often employ amplicon-based sequencing followed by mapping of sequencing reads to a closely related reference genome, facilitating the assembly of those reads into a consensus genome. For bacteria, with their genome content variation and extensive within-species diversity, there are some nuances to genome assembly and analysis that should be considered. Below we will describe the main approaches for bacterial genome assembly, and how each approach may impact downstream inference of relatedness.

The two main approaches to bacterial genome assembly are **reference-based assembly** and *de novo* **assembly**. For reference-based assembly, sequencing reads are mapped to a closely related reference genome, mirroring the amplicon-based approach we commonly use for viruses. *De novo* assembly, also referred to as reference-free assembly, uses information from overlapping portions of the sequencing reads to piece the genome together like a jigsaw puzzle. While reference-based assembly will result in a consensus sequence with the same length as the reference genome, *de novo* assembly will result in several (anywhere from tens to hundreds) of **contigs** – a contiguous sequence derived from multiple overlapping sequencing reads and representing a large portion of the bacterial genome. Contigs inferred during *de novo* assembly are then annotated using tools like Prokka [31] or NCBI's Prokaryotic Genome Annotation Pipeline (PGAP) [32]. Each approach has associated advantages and limitations, which we will discuss, and in practice, the two are often used in tandem.

The main disadvantage of reference-based assembly is that reference selection can significantly impact downstream analysis. Most notably, you can only identify variation in regions of the genome that are shared between the reference and the sequenced bacterial isolates. If the strain of interest possesses mobile genetic elements or genes acquired through HGT that are not present in the reference, then the ability to identify SNPs in those loci is lost. Overall, the more diverged the reference, the less "callable" the genome. If the reference is diverged to the point that there is an appreciable amount of mutations, insertions and deletions (often shortened as indels), or genome rearrangements, then read mapping could be adversely affected. This could result in an artificial reduction in SNP distances amongst your bacterial isolates of interest, which would likely impact the inference of epidemiological linkage. A good metric to assess the appropriateness of the reference is to measure the percentage of reads mapping to the reference during the mapping step of genome assembly. If the reference is a close match, you should expect to see 96%–99% of sequencing reads mapping to the reference.

With *de novo* assembly, the organization of the genome is lost, that is, the contigs are not ordered in the way they are found in the bacterial chromosome. As a result, contigs from separate isolates cannot simply be aligned to each other to identify SNPs. The solution to this issue is to first identify all of the coding sequences in the *de novo* genome assembly through the process of genome annotation. Afterwards, the annotated sequences are used as the input for pangenome analysis programs that then identify and cluster all of the orthologous coding sequences; these are referred to as **clusters of orthologous groups** or COGs. Subsequently, each COG is individually aligned, which is a far easier computational feat than attempting to align complete bacterial genomes. Afterwards, the core and accessory genome of a set of sequences is defined based on the presence and absence of COGs in the dataset. The alignments for COGs comprising the core genome can then be concatenated into a core genome alignment. It's worth noting that these alignments do not include intergenic regions of the bacterial chromosome, which contain the biologically relevant gene promoters. In addition, information about the distance between COGs is lost. Further, while core genome alignments only contain a portion of the total genome content, file sizes can still be very large, especially for datasets including hundreds of isolates. To address this, users often extract only the variable sites, creating a core genome SNP alignment. The SNP alignment is then used for downstream phylogenetic analysis.

8.5.1   Considerations for Long-Read Sequencing Data

Sequencing platforms can broadly be categorised into those that produce "short" and "long" reads. Short read platforms like those from Illumina produce reads ranging from 50 to 350 base pairs while long read sequencers such as those from PacBio and Oxford Nanopore Technology (ONT) generate reads from 1,000 to longer than 10,000 base pairs in length. A key difference is that while short read sequencers produce uniform read lengths, long read sequencers result in a distribution of read lengths. In addition, long read sequencers have lower per-base read accuracy than short read sequencers. Notwithstanding, the popularity of long-read sequencers has grown significantly as platforms like those from

ONT offer flexibility in sequencing throughput. For example, there is no set time for the duration of an ONT sequencing run – one can run it until sufficient data has been generated and the flow cell can be reused as long as there are remaining active pores. In addition, base calling occurs in real time, which offers a number of potential applications for diagnostics and genomic epidemiology. An attractive feature of long reads is that they can be used to more accurately determine genome organization. However, people often cite the lower accuracy as a downside of the technology. To address the lower accuracy compared to short reads, methods have been developed to combine short and long read data into hybrid genome assemblies. For hybrid assembly, a short-read-only *de novo* assembly is first generated, then the long reads are used to order (scaffold) the contigs. The short reads are further used to correct any errors in the assembly by iteratively mapping the reads to the draft genome in a process called "polishing". It is not uncommon for this approach to result in a fully circularised "closed" genome. While the ability of long read sequencing data to create closed genomes is certainly impressive, at present, it likely provides a negligible advantage for most goals of genomic epidemiology analysis. As the accuracy of long read sequencers improves, long-read only *de novo* assemblies may present an alternative for regular use. At the time of writing, ONT had released updated flow cells and corresponding library preparation chemistry that were capable of >99% accuracy.

8.6 A PRACTICAL WORKFLOW FOR BACTERIAL GENOMIC EPIDEMIOLOGY

Here, we will provide an overview of a generic workflow for bacterial genomic epidemiology. While the steps may vary based on the bacterial species of interest, the general process is overall consistent. We urge the user to first investigate the genomics of the microbe. This should include an understanding of the basic genome structure (e.g., number of chromosomes, genome length, presence of plasmids, average number of coding sequences, and GC content), whether the species is known to be naturally competent (i.e., able to undergo homologous recombination), and the basics of population structure and molecular typing. We will assume that short-read sequencing data has been generated for a set of isolates, as this is currently the most common platform in public health labs, and that the necessary quality control steps have been performed.

The workflow usually begins with a *de novo* assembly of all the samples followed by genome annotation. A quality control step at this point involves assessing the statistics from the annotation process to identify the number of assembled contigs, total genome length, GC content, and the number of coding sequences (CDSs) annotated. By plotting the statistics, for example using a basic histogram, the user can flag outliers, which should either be further examined or excluded from the subsequent analysis. *De novo* assemblies can then be used as inputs for a number of tools to confirm the species, identify genotypic antibiotic resistance and virulence determinants, assign a sequence type, and genotypically assign serotypes or serogroups, if relevant to the species of interest.

After this step, the annotation files are used by pangenome analysis tools to determine the core genome. As detailed above, this process will result in a core-genome SNP alignment that can be used to produce a SNP distance matrix or phylogeny. As a quick technical note about best practices, users should use phylogenetic algorithms that take into account SNP alignments. These algorithms usually include some form of ascertainment bias correction that corrects for the "missing" nucleotides in the alignment that were removed because they did not show polymorphism. In addition, users should be reminded that if they are working with a recombining species, this alignment will also include SNPs introduced through recombination, unless recombination detection and masking was previously performed.

At this point in the analysis, the phylogeny should be visualised in conjunction with any typing, antibiotic resistance, or epidemiological data that are relevant. A common practice is to annotate the phylogeny with the typing information (e.g., MLST) and to use a heatmap to illustrate the epidemiological data associated with each sample. If the goal of the analysis is to identify samples that may be related through recent transmission, then this step can be used to identify samples that can be excluded (i.e., "ruled-out") from the putative transmission clusters. This will often focus the analysis on a subset of samples. Since the inclusion of "outliers" reduces the core genome size, the terminal branch lengths on the lineage or clade of interest will be artificially lower, as discussed above. To address this, the samples excluded from the putative outbreak cluster, should be removed from the dataset and the pangenome analysis repeated. This secondary analysis will yield more accurate SNP distances that can be used to infer transmission events.

The workflow above avoided the use of reference-based genome assembly and can be used for a variety of applications. However, if the species is known to experience homologous recombination, then the removal of genomic regions that have been impacted by recent recombination events may be necessary. If this is the case, then the above workflow should still be used to identify the clade or cluster of interest. An appropriate reference genome can then be identified using a number of available tools. Thereafter, the sequence reads from each sample in the subset of interest can be mapped to the reference and a full length multiple sequence alignment can be obtained. Thankfully, over the last decade, bioinformatics tools have significantly advanced and readily available tools like Torsten Seemann's Snippy (https://github.com/tseemann/snippy) can perform much of this analysis. Once created, the multiple sequence alignment will be used as an input for recombination detection tools. The reference based assembled and full length alignment is needed because most recombination analysis tools use the distance between loci to detect evidence for recombination. As noted, these programs will "censor" areas of the genome that have been impacted by recombination and produce a "recombination free" alignment. In addition, they will output several statistics including the ratio of SNPs introduced through recombination compared to mutation (r/m).

8.7 INTRAHOST DIVERSITY AND IMPLICATIONS FOR INFERRING TRANSMISSION

As we move towards a discussion about how to infer the relatedness among bacterial isolates using whole genome sequencing data, we must first consider an important aspect of microbial transmission and evolution. A bacterial population present during an episode of asymptomatic carriage or disease is just that, a population.

While most genomic epidemiology studies employ the sequencing of a single isolate, this practice is largely the result of the dogma of classical microbiology techniques combined with computational and cost constraints. Recent studies have described the amount of microbial diversity that can exist in an individual, referred to as **intra-host** or **within-host diversity**, which is not captured by sampling only a single isolate. This topic won't be explored in its entirety here, but a brief overview is needed to understand how intrahost diversity may impact the investigation of transmission dynamics during an outbreak.

When a bacterium is acquired, it begins to replicate. As described previously, uncorrected errors occurring during replication can result in the ongoing accrual of mutations and indels. In addition, bacteria often come into contact with other microbial species and sometimes other members of the same species. This provides the opportunity for acquisition of mobile genetic elements such as phages or plasmids, or in the case of naturally competent bacteria, to undergo homologous recombination using DNA taken up from the environment. The combined effects of these processes results in appreciable diversity within the bacterial population. This diversity may extend beyond nucleotide diversity and include gene content variation.

The amount of diversity that accrues is heavily dependent on the bacterial species and whether the acquisition results in carriage or disease. For carried bacteria, the duration and site of carriage can also impact the breadth of diversity. Sites may include the lower or upper respiratory tract, skin, or gastrointestinal tract, and carriage at multiple anatomic sites is not uncommon. The longer a bacterial population is carried, the more opportunity there is to come in contact with other bacteria and to also accumulate mutations. Put simply, increasing carriage duration results in a larger and a more diverse population. For bacterial acquisitions that result in infection, there is often less opportunity to accumulate diversity since this process usually occurs over shorter timescales.

One potential outcome of acquiring a pathogen is subsequently transmitting that pathogen on. During a transmission event, only a subsample of the total diversity present in a host is transferred. The amount of diversity transferred is referred to as the **transmission bottleneck size** and is dependent on the mode of transmission and the amount of diversity that was present on the primary host. For example, transmission of a respiratory pathogen via a fomite may have a smaller transmission bottleneck size than direct respiratory droplet transmission. Each transmission event can transfer a different random subsample of diversity. Because of this, sampling a single bacterial isolate from each individual in a transmission chain that involves one person transmitting to multiple recipients will more than likely result in difficulty inferring the direction of transmission. This is a current limitation in methodological approaches to inferring transmission. Below we will discuss potential future solutions. For a more complete discussion of within host evolution, Didelot and colleagues [33] provide an excellent review.

8.8   INVESTIGATION OF TRANSMISSION

The application of bacterial whole genome sequencing to investigating transmission has provided an unprecedented level of resolution beyond what was previously capable with traditional molecular typing approaches. Still, the process of inferring transmission using genomic data remains complex and is often debated. The approaches, ordered by increasing complexity, are 1) SNP-distance based methods, 2) phylogenetic methods, and 3) modeling based methods. Before diving into the details of each, the one constant across all approaches is that epidemiological data must be taken into consideration when inferring transmission. No pathogen-centric method, sequencing or otherwise, will ever replace the need for a classic shoe-leather epidemiology investigation. Another truism is that it is far easier to *rule-out* a case using pathogen genomic data than it is to *rule-in*.

SNP-distance based methods are the easiest and most simplistic method for ruling-in and -out cases from a putative transmission cluster. This method uses a SNP cutoff to define isolates from cases likely related by recent transmission. These cutoffs can be set *a priori* based on previous outbreak investigations and knowledge of the organism or empirically based on the data collected during the investigation at hand. For example, SNP cutoffs for *S. aureus* have long been held as ranging from 30 to 40 SNPs between related isolates. An extensive study of *S. aureus* transmission modeling later found that a cutoff of 39 SNPs [34] yielded a high sensitivity and specificity for inferring transmission. Of course, since genome size, recombination rates, and evolutionary rates vary greatly among bacterial species, one single cutoff is not suitable for all.

If an analyst obtains a general sample of bacterial genomes from a species of interest as well as those that are suspected as being from an outbreak, then they can empirically define a SNP cutoff. To do this, one obtains a pairwise SNP distance matrix and then plots the distribution of SNP distances using a histogram (see Figure 8.1 for an example). This distribution will be bi- or tri-modal (having two to three peaks). The furthest right peak will illustrate the SNP distances between distantly related members of the species, the middle peak will represent the members of a lineage or clone of that species, and the furthest left peak will be the genomes that are the most closely related – and most likely to be linked by recent transmission. Therefore, you could set your SNP cutoff to fall between the first and second peak (looking from left to right).

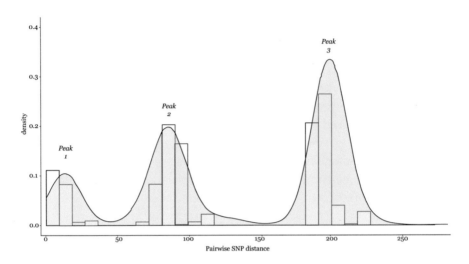

Figure 8.1   A histogram of theoretical SNP distances between various bacterial core genome sequences. In this case, the distribution is trimodal. A small number of sequences cluster together with very few SNP differences, likely signalling relatedness, while other groups of isolates are somewhat more different from each other.

When combined with epidemiological data (e.g., overlapping lengths of stay in a hospital, consumption of the same food item, or sexual contact), this may be all you need to resolve the outbreak. Increasingly, methods such as those from Duval and colleagues [35] facilitate the estimation of outbreak specific SNP thresholds, accounting for organism and sample-specific evolutionary rates, and variation in outbreak duration.

Phylogenetic methods are more complex and not as easy to interpret, a classical example of increased data richness also yielding greater nuance and clearer underlying uncertainty. A staunch phylogeneticist would argue that you are "throwing away" important data by only considering the number of SNP differences and not incorporating our understanding of molecular evolution. Therefore, one should use a phylogenetic tree inferred from the core genome alignment to investigate transmission. The catch is that one's interpretation of the tree is always caveated since the phylogenetic tree does not always match the transmission tree. In addition, evolutionary distances used when displaying a tree may not be readily interpretable for analysts new to reading trees. Nevertheless, phylogenetic-based inferences of transmission are likely to provide the best resolution when inferring transmission patterns. As an extension of the current phylogenetic approaches, methods are being developed to incorporate sequencing from high density sampling (i.e., sequencing up to eight to ten isolates from a single case) or high depth sequencing of the entire bacterial population, referred to as deep sequencing. These approaches address the issue of intrahost diversity and promise the ability to infer the directionality of transmission [36]. The goal here is to identify whether the "cloud of diversity" characterizing one case's bacterial population is nested within the genetic diversity of another individual's bacterial population. The more the diversity overlaps, the more confidence there is that the populations are directly related through transmission [37]. When a temporal component is incorporated into the analysis, then the directionality of transmission can be inferred.

The term "modeling based methods" is used here to describe computational approaches that incorporate epidemiological and pathogen genomic data to jointly infer transmission, such as TransPhylo [38,39]. Epidemiological data could include dates of disease onset and laboratory report dates as well as estimates of the infectious period. Values such as the transmission bottleneck size can also be estimated. An advantage of this approach is that it provides a probability of transmission between two cases. This is particularly appealing because public health decision makers often want to know how confident analysts are in their findings when a simple "yes/no" or "rule-in/out" answer is not possible (which it rarely ever is). The disadvantage is that these methods are largely accessible to only advanced analysts. However, as these methods are improved upon, they could eventually be incorporated into popular genomic epidemiology graphical user interfaces. In addition, as sampling strategies evolve, the use of deep sequencing or high-density sampling could significantly advance our ability to determine the true transmission tree [34].

8.9 CONCLUSIONS

Taken together, due to the complexity of their genomes and unique evolutionary dynamics, the application of genomic epidemiology to bacterial species requires specialised knowledge distinct from viruses. However, much of the methodological approaches are consistent when investigating suspected viral or bacterial outbreaks. There are countless success stories in bacterial genomic epidemiology, and as a result, genomic surveillance of select bacterial pathogens is routine in many jurisdictions. Large bacterial genomic databases are now available for several species, which can be used to inform our knowledge about diversity and distribution of the population. Again, one must consider the nuances of bacterial evolution when investigating transmission. This includes an understanding of 1) the epidemiology of the pathogen, 2) recombination and mutation and how they shape population structure with the help of selection and drift, 3) genome content variation and pangenome analysis, and 4) intrahost evolution. With ever-evolving computational tools, bacterial genomic epidemiology is becoming more and more commonplace. The hope is that this will ultimately lead to earlier identification of emerging threats and rapid case investigation to mitigate spread.

References

[1] Gregory L Armstrong, Duncan R MacCannell, Jill Taylor, Heather A Carleton, Elizabeth B Neuhaus, Richard S Bradbury, James E Posey, and Marta Gwinn. Pathogen genomics in public health. *New England Journal of Medicine*, 381(26):2569–2580, 2019.

[2] Emma B Hodcroft, Moira Zuber, Sarah Nadeau, Timothy G Vaughan, Katharine HD Crawford, Christian L Althaus, Martina L Reichmuth, John E Bowen, Alexandra C Walls, Davide Corti, et al. Spread of a SARS-CoV-2 variant through Europe in the summer of 2020. *Nature*, 595(7869):707–712, 2021.

[3] Nicholas G Davies, Sam Abbott, Rosanna C Barnard, Christopher I Jarvis, Adam J Kucharski, James D Munday, Carl AB Pearson, Timothy W Russell, Damien C Tully, Alex D Washburne, et al. Estimated transmissibility and impact of SARS-CoV-2 lineage B. 1.1. 7 in England. *Science*, 372(6538):eabg3055, 2021.

[4] Erik Volz, Swapnil Mishra, Meera Chand, Jeffrey C Barrett, Robert Johnson, Lily Geidelberg, Wes R Hinsley, Daniel J Laydon, Gavin Dabrera, Áine O'Toole, et al. Assessing transmissibility of SARS-CoV-2 lineage B. 1.1. 7 in England. *Nature*, 593(7858):266–269, 2021.

[5] Michael Worobey, Marlea Gemmel, Dirk E Teuwen, Tamara Haselkorn, Kevin Kunstman, Michael Bunce, Jean-Jacques Muyembe, Jean-Marie M Kabongo, Raphaël M Kalengayi, Eric Van Marck, et al. Direct evidence of extensive diversity of HIV-1 in Kinshasa by 1960. *Nature*, 455(7213):661–664, 2008.

[6] Ben Krause-Kyora, Julian Susat, Felix M Key, Denise Kühnert, Esther Bosse, Alexander Immel, Christoph Rinne, Sabin-Christin Kornell, Diego Yepes, Sören Franzenburg, et al. Neolithic and

medieval virus genomes reveal complex evolution of hepatitis B. *Elife*, 7:e36666, 2018.

[7] Gytis Dudas, Luiz Max Carvalho, Trevor Bedford, Andrew J Tatem, Guy Baele, Nuno R Faria, Daniel J Park, Jason T Ladner, Armando Arias, Danny Asogun, et al. Virus genomes reveal factors that spread and sustained the Ebola epidemic. *Nature*, 544(7650):309–315, 2017.

[8] John W Drake. Rates of spontaneous mutation among RNA viruses. *Proceedings of the National Academy of Sciences*, 90(9):4171–4175, 1993.

[9] José M Malpica, Aurora Fraile, Ignacio Moreno, Clara I Obies, John W Drake, and Fernando García-Arenal. The rate and character of spontaneous mutation in an RNA virus. *Genetics*, 162(4):1505–1511, 2002.

[10] Edward C Holmes. *The evolution and emergence of RNA viruses*. Oxford University Press, 2009.

[11] John T McCrone and Adam S Lauring. Genetic bottlenecks in intraspecies virus transmission. *Current opinion in virology*, 28:20–25, 2018.

[12] Edward C Holmes, Gytis Dudas, Andrew Rambaut, and Kristian G Andersen. The evolution of Ebola virus: Insights from the 2013–2016 epidemic. *Nature*, 538(7624):193–200, 2016.

[13] Suzanne E Mate, Jeffrey R Kugelman, Tolbert G Nyenswah, Jason T Ladner, Michael R Wiley, Thierry Cordier-Lassalle, Athalia Christie, Gary P Schroth, Stephen M Gross, Gloria J Davies-Wayne, et al. Molecular evidence of sexual transmission of Ebola virus. *New England Journal of Medicine*, 373(25):2448–2454, 2015.

[14] Placide Mbala-Kingebeni, Catherine Pratt, Mbusa Mutafali-Ruffin, Matthias G Pauthner, Faustin Bile, Antoine Nkuba-Ndaye, Allison Black, Eddy Kinganda-Lusamaki, Martin Faye, Amuri Aziza, et al. Ebola virus transmission initiated by relapse of systemic Ebola virus disease. *New England Journal of Medicine*, 384(13):1240–1247, 2021.

[15] Nuno R Faria, Joshua Quick, Ingra M Claro, Julien Theze, Jacqueline G de Jesus, Marta Giovanetti, Moritz UG Kraemer, Sarah C Hill, Allison Black, Antonio C da Costa, et al. Establishment and cryptic transmission of Zika virus in Brazil and the Americas. *Nature*, 546(7658):406–410, 2017.

[16] Yatish Turakhia, Bryan Thornlow, Angie S Hinrichs, Nicola De Maio, Landen Gozashti, Robert Lanfear, David Haussler, and Russell Corbett-Detig. Ultrafast Sample placement on Existing tRees (UShER) enables real-time phylogenetics for the SARS-CoV-2 pandemic. *Nature Genetics*, 53(6):809–816, 2021.

[17] Ivan Aksamentov, Cornelius Roemer, Emma B Hodcroft, and Richard A Neher. Nextclade: clade assignment, mutation calling and quality control for viral genomes. *Journal of open source software*, 6(67):3773, 2021.

[18] James Hadfield, Colin Megill, Sidney M Bell, John Huddleston, Barney Potter, Charlton Callender, Pavel Sagulenko, Trevor Bedford, and Richard A Neher. Nextstrain: real-time tracking of pathogen evolution. *Bioinformatics*, 34(23):4121–4123, 2018.

[19] Lam-Tung Nguyen, Heiko A Schmidt, Arndt Von Haeseler, and Bui Quang Minh. IQ-TREE: a fast and effective stochastic algorithm for estimating maximum-likelihood phylogenies. *Molecular biology and evolution*, 32(1):268–274, 2015.

[20] Alexey M Kozlov, Diego Darriba, Tomáš Flouri, Benoit Morel, and Alexandros Stamatakis. RAxML-NG: a fast, scalable and user-friendly tool for maximum likelihood phylogenetic inference. *Bioinformatics*, 35(21):4453–4455, 2019.

[21] Morgan N Price, Paramvir S Dehal, and Adam P Arkin. FastTree 2–approximately maximum-likelihood trees for large alignments. *PloS one*, 5(3):e9490, 2010.

[22] Alexei J Drummond and Andrew Rambaut. BEAST: Bayesian evolutionary analysis by sampling trees. *BMC evolutionary biology*, 7(1):1–8, 2007.

[23] Ben Jackson, Maciej F Boni, Matthew J Bull, Amy Colleran, Rachel M Colquhoun, Alistair C Darby, Sam Haldenby, Verity Hill, Anita Lucaci, John T McCrone, et al. Generation and transmission

of interlineage recombinants in the SARS-CoV-2 pandemic. *Cell*, 184(20):5179–5188, 2021.

[24] Niranjan Nagarajan and Carl Kingsford. GiRaF: robust, computational identification of influenza reassortments via graph mining. *Nucleic acids research*, 39(6):e34–e34, 2011.

[25] Pierre Barrat-Charlaix, Timothy G Vaughan, and Richard A Neher. Treeknit: Inferring ancestral reassortment graphs of influenza viruses. *PLoS Computational Biology*, 18(8):e1010394, 2022.

[26] Nicola F Müller, Ugnė Stolz, Gytis Dudas, Tanja Stadler, and Timothy G Vaughan. Bayesian inference of reassortment networks reveals fitness benefits of reassortment in human influenza viruses. *Proceedings of the National Academy of Sciences*, 117(29):17104–17111, 2020.

[27] Remco Bouckaert, Joseph Heled, Denise Kühnert, Tim Vaughan, Chieh-Hsi Wu, Dong Xie, Marc A Suchard, Andrew Rambaut, and Alexei J Drummond. BEAST 2: a software platform for Bayesian evolutionary analysis. *PLoS computational biology*, 10(4):e1003537, 2014.

[28] Sebastian Duchêne, Kathryn E. Holt, François-Xavier Weill, Simon Le Hello, Jane Hawkey, David J. Edwards, Mathieu Fourment, and Edward C. Holmes. Genome-scale rates of evolutionary change in bacteria. *Microbial Genomics*, 2(11), 2016.

[29] Edward J Feil and Brian G Spratt. Recombination and the population structures of bacterial pathogens. *Annual Reviews in Microbiology*, 55(1):561–590, 2001.

[30] Valeria Mateo-Estrada, José Luis Fernández-Vázquez, Julia Moreno-Manjón, Ismael L Hernández-González, Eduardo Rodríguez-Noriega, Rayo Morfín-Otero, María Dolores Alcántar-Curiel, and Santiago Castillo-Ramírez. Accessory genomic epidemiology of cocirculating Acinetobacter baumannii clones. *Msystems*, 6(4):e00626–21, 2021.

[31] Torsten Seemann. Prokka: rapid prokaryotic genome annotation. *Bioinformatics*, 30(14):2068–2069, 2014.

[32] Tatiana Tatusova, Michael DiCuccio, Azat Badretdin, Vyacheslav Chetvernin, Eric P Nawrocki, Leonid Zaslavsky, Alexandre Lomsadze, Kim D Pruitt, Mark Borodovsky, and James Ostell. NCBI

prokaryotic genome annotation pipeline. *Nucleic acids research*, 44(14):6614–6624, 2016.

[33] Xavier Didelot, A Sarah Walker, Tim E Peto, Derrick W Crook, and Daniel J Wilson. Within-host evolution of bacterial pathogens. *Nature Reviews Microbiology*, 14(3):150–162, 2016.

[34] Matthew D Hall, Matthew TG Holden, Pramot Srisomang, Weera Mahavanakul, Vanaporn Wuthiekanun, Direk Limmathurotsakul, Kay Fountain, Julian Parkhill, Emma K Nickerson, Sharon J Peacock, et al. Improved characterisation of MRSA transmission using within-host bacterial sequence diversity. *Elife*, 8:e46402, 2019.

[35] Audrey Duval, Lulla Opatowski, and Sylvain Brisse. Defining genomic epidemiology thresholds for common-source bacterial outbreaks: a modelling study. *The Lancet Microbe*, 4(5):e349–e357, 2023.

[36] Chris Wymant, Matthew Hall, Oliver Ratmann, David Bonsall, Tanya Golubchik, Mariateresa de Cesare, Astrid Gall, Marion Cornelissen, Christophe Fraser, The Maela Pneumococcal Collaboration STOP-HCV Consortium, and The BEEHIVE Collaboration. PHYLOSCANNER: inferring transmission from within-and between-host pathogen genetic diversity. *Molecular biology and evolution*, 35(3):719–733, 2018.

[37] Steven YC Tong, Matthew TG Holden, Emma K Nickerson, Ben S Cooper, Claudio U Köser, Anne Cori, Thibaut Jombart, Simon Cauchemez, Christophe Fraser, Vanaporn Wuthiekanun, et al. Genome sequencing defines phylogeny and spread of methicillin-resistant Staphylococcus aureus in a high transmission setting. *Genome research*, 25(1):111–118, 2015.

[38] Xavier Didelot, Michelle Kendall, Yuanwei Xu, Peter J White, and Noel McCarthy. Genomic epidemiology analysis of infectious disease outbreaks using TransPhylo. *Current protocols*, 1(2):e60, 2021.

[39] Xavier Didelot, Christophe Fraser, Jennifer Gardy, and Caroline Colijn. Genomic infectious disease epidemiology in partially sampled and ongoing outbreaks. *Molecular biology and evolution*, 34(4):997–1007, 2017.

Index

Printed in the United States
by Baker & Taylor Publisher Services